Rebecca Kohler

Membrane protein insertion

Rebecca Kohler

Membrane protein insertion

A comparison of two protein insertion systems

Südwestdeutscher Verlag für Hochschulschriften

Impressum/Imprint (nur für Deutschland/only for Germany)
Bibliografische Information der Deutschen Nationalbibliothek: Die Deutsche Nationalbibliothek verzeichnet diese Publikation in der Deutschen Nationalbibliografie; detaillierte bibliografische Daten sind im Internet über http://dnb.d-nb.de abrufbar.
Alle in diesem Buch genannten Marken und Produktnamen unterliegen warenzeichen-, marken- oder patentrechtlichem Schutz bzw. sind Warenzeichen oder eingetragene Warenzeichen der jeweiligen Inhaber. Die Wiedergabe von Marken, Produktnamen, Gebrauchsnamen, Handelsnamen, Warenbezeichnungen u.s.w. in diesem Werk berechtigt auch ohne besondere Kennzeichnung nicht zu der Annahme, dass solche Namen im Sinne der Warenzeichen- und Markenschutzgesetzgebung als frei zu betrachten wären und daher von jedermann benutzt werden dürften.

Coverbild: www.ingimage.com

Verlag: Südwestdeutscher Verlag für Hochschulschriften GmbH & Co. KG
Heinrich-Böcking-Str. 6-8, 66121 Saarbrücken, Deutschland
Telefon +49 681 37 20 271-1, Telefax +49 681 37 20 271-0
Email: info@svh-verlag.de

Approved by: Zürich, ETH, Diss., 2010

Herstellung in Deutschland:
Schaltungsdienst Lange o.H.G., Berlin
Books on Demand GmbH, Norderstedt
Reha GmbH, Saarbrücken
Amazon Distribution GmbH, Leipzig
ISBN: 978-3-8381-3096-5

Imprint (only for USA, GB)
Bibliographic information published by the Deutsche Nationalbibliothek: The Deutsche Nationalbibliothek lists this publication in the Deutsche Nationalbibliografie; detailed bibliographic data are available in the Internet at http://dnb.d-nb.de.
Any brand names and product names mentioned in this book are subject to trademark, brand or patent protection and are trademarks or registered trademarks of their respective holders. The use of brand names, product names, common names, trade names, product descriptions etc. even without a particular marking in this works is in no way to be construed to mean that such names may be regarded as unrestricted in respect of trademark and brand protection legislation and could thus be used by anyone.

Cover image: www.ingimage.com

Publisher: Südwestdeutscher Verlag für Hochschulschriften GmbH & Co. KG
Heinrich-Böcking-Str. 6-8, 66121 Saarbrücken, Germany
Phone +49 681 37 20 271-1, Fax +49 681 37 20 271-0
Email: info@svh-verlag.de

Printed in the U.S.A.
Printed in the U.K. by (see last page)
ISBN: 978-3-8381-3096-5

Copyright © 2012 by the author and Südwestdeutscher Verlag für Hochschulschriften GmbH & Co. KG and licensors
All rights reserved. Saarbrücken 2012

Summary 5

Zusammenfassung 6

1 Introduction 9

   1.1 The Bacterial Ribosome 9

   1.2 Insertion of Nascent Membrane Proteins into the Membrane 11

      1.2.1 The SRP-SecYEG insertion pathway 11

      1.2.2 The role of the YidC/Oxa1/Alb3 protein family in membrane insertion 12

      1.2.3 Function of YidC in Sec-dependent membrane insertion 13

      1.2.4 The Sec-independent YidC pathway of membrane protein insertion 13

   1.3 Structural Information on YidC 14

   1.4 The Mitochondrial Oxa1 Protein 15

   1.5 Aim of this Project 16

2 Materials 18

   2.2 Chemicals 18

   2.2 Enzymes 19

   2.3 DNA and Protein Molecular Weight Markers 20

   2.4 Buffers and Solutions 21

      2.4.1 Antibiotics and inducers 21

      2.4.2 Electrophoresis buffers 21

      2.4.3 Buffers for protein purification 22

      2.4.4 Other buffers and solutions 23

      2.4.5 Microbial growth media 23

      2.4.6 *E. coli* strains 24

      2.4.7 Commercial kits 24

      2.4.8 Materials for cryo-EM and structure calculation 24

3 Methods 26

   3.1 DNA Methods 26

      3.1.1 Preparation of genomic DNA from *E. coli* 26

      3.1.2 Preparation of plasmid DNA 26

      3.1.3 Agarose gel electrophoresis 26

      3.1.4 Polymerase chain reaction 27

      3.1.5 Restriction digests 27

      3.1.6 TOPO® cloning (Invitrogen) 28

      3.1.7 Ligation of DNA fragments 28

- 3.1.8 Transformation of electro-competent *E. coli* cells 28
- 3.1.9 DNA sequencing 28
- 3.1.10 Expression of YidC and Oxa1 constructs 28
- 3.1.11 Membrane preparation of YidC and Oxa1 constructs 29
- 3.1.12 Purification of YidC constructs 29
- 3.1.13 Purification of Oxa1 30
- 3.1.14 Preparation of membrane free cell extract 30
- 3.1.14 Preparation of *E. coli* 70S ribosomes 30
- 3.1.15 Preparation of *E. coli* ribosome nascent chain complexes (RNCs) 30
- 3.1.16 Sedimentation assays 31
- 3.1.17 Analysis of oligomeric state by native PAGE 31
- 3.1.18 PICUP crosslinking assays 31
- 3.1.19 Disulfide crosslinking assays 32
- 3.1.20 Sample preparation and data acquisition by cryo-electron microscopy 32
- 3.1.21 Data processing and 3D reconstruction 32
- 3.1.22 Generation of masks 36
- 3.1.23 Extrapolation of the resolution value to the full data set 36

4 Results 37
- 4.1 Cloning of YidC and Oxa1 Constructs 37
- 4.2 Cloning of F_0c Ribosome Nascent Chain Constructs 38
- 4.3 Preparation of Ribosome Nascent Chain Constructs 38
- 4.4 Purification of His-tagged YidC and Oxa1 Proteins 39
- 4.5 The Cytoplasmic C-terminus of YidC Binds to the Ribosome 40
- 4.6 Oxa1 Interacts with the Bacterial Ribosome 42
- 4.7 EM Data Collection and Processing 43
- 4.8 3D Reconstruction of YidC Bound to the Translating Ribosome 45
- 4.9 The RNC-Oxa1 Complex 49
- 4.10 Dimers of YidC and Oxa1 Bind to the Ribosome 50
  - 4.10.1 Analysis of the oligomeric state of YidC in solution by BN-PAGE 50
  - 4.10.2 Analysis of the oligomeric state of YidC on the ribosome by $Ru(bpy)_3$ crosslinking 51
- 4.11 Cysteine Crosslinking of YidC and Oxa1 Dimers 53

5 Discussion 56

6 Appendix 63
- 6.1 DNA and Protein Sequences of YidC and Oxa1 Constructs 63

7 References 69
Glossary 74

Parts of this thesis have been published in

Kohler, R., D. Boehringer, et al. (2009). "YidC and Oxa1 form dimeric insertion pores on the translating ribosome." Mol Cell **34**(3): 344-53.

SUMMARY

Membrane proteins play important roles in diverse cellular processes such as energy conversion, cellular transport, signal transduction and cell division. Like all other proteins, membrane proteins are synthesized by the ribosome. Membrane proteins have a high content of hydrophobic amino acids and thus risk aggregation in aqueous environments like the cytosol. Therefore, most bacterial inner membrane proteins are co-translationally inserted into the membrane, thereby avoiding the unfavorable cytosolic environment.

In bacteria, most nascent inner membrane proteins are recognized by the ubiquitous signal recognition particle (SRP) that subsequently targets its bound substrate to SecYEG. The Sec-translocon forms a protein conducting channel in the membrane from which inserted transmembrane helices can exit laterally into the surrounding membrane bilayer. In this "Sec-dependent" pathway of membrane insertion, the *Escherichia coli* inner membrane protein YidC performs an accessory role, probably by facilitating substrate exit from the translocation channel into the membrane.

In addition to the canonical SRP-Sec pathway of membrane protein insertion, a more recently discovered pathway exists for the membrane insertion of a set of rather short and hydrophobic transmembrane substrates: The YidC- or "Sec-independent" pathway of membrane insertion. In bacteria, mitochondria and chloroplasts, members of the YidC/Oxa1/Alb3 family of membrane proteins facilitate the insertion and assembly of membrane proteins.

In this thesis, evidence for a direct interaction between *E. coli* YidC and the translating as well as non-translating ribosome is provided. A similar interaction could be confirmed for the *Saccharomyces cerevisiae* Oxa1 (a YidC homolog) and *E. coli* ribosomes. The structures of both *E. coli* YidC and *S. cerevisiae* Oxa1 bound to *E. coli* ribosome nascent chain complexes determined by cryo-electron microscopy are presented. In the cryo-EM densities, dimers of YidC and Oxa1 are localized above the exit of the ribosomal tunnel. In addition to this, crosslinking experiments show that the ribosome specifically stabilizes the dimeric state of both YidC and Oxa1. Functionally important and conserved transmembrane helices of YidC and Oxa1 were localized at the dimer interface by cysteine crosslinking. Interestingly, both Oxa1 and YidC dimers contact the ribosome at ribosomal protein L23 and conserved rRNA helices 59 and 24 similarly to the contacts observed for the non-homologous SecYEG translocon. On the basis of these results, we suggest that dimers of the YidC and Oxa1 proteins form insertion pores and share a common overall architecture of a protein translocation/insertion pore with the SecY monomer.

ZUSAMMENFASSUNG

Membranproteine spielen in vielen unterschiedlichen zellulären Prozessen eine wichtige Rolle: In der biochemischen Umwandlung von Energie, im zellulären Transport, in der Signaltransduktion und der Zellteilung. Wie alle anderen Proteine werden Membranproteine am Ribosom synthetisiert. Membranproteine bestehen im Allgemeinen aus einem hohen Anteil hydrophober Aminosäuren und laufen deshalb Gefahr, in wässriger Umgebung wie dem Zytosol zu aggregieren. Aus diesem Grund werden die meisten Membranproteine der inneren Bakterienmembran ko-translational in die Membran insertiert, wobei ein Kontakt der naszierenden Proteinkette mit der für sie ungünstigen zytosolischen Umgebung vermieden wird.

In Bakterien werden die meisten Proteine der inneren Membran vom „signal recognition particle" (SRP) erkannt, welches sein gebundenes Substrat anschliessend zu SecYEG dirigiert. Das Sec-Translocon bildet einen Kanal in der Membran, den die bereits insertierten Transmembranhelices seitlich verlassen können, um in die umgebende Membran zu gelangen. In diesem „Sec-abhängigen" Membraninsertionsweg spielt das YidC-Protein, das in der inneren Membran von *Escherichia Coli* vorkommt, eine Hilfsrolle; vermutlich, indem es den Austritt des Substrats aus dem Translokationskanal in die Membran vereinfacht.

Zusätzlich zu dem bereits wohletablierten SRP-Sec-Membraninsertionsweg existiert auch ein alternativer Weg für die Membraninsertion einer Reihe von eher kurzen und hydrophoben Transmembransubstraten, der erst kürzlich entdeckt wurde: Der YidC- oder „Sec-unabhängige" Membraninsertionsweg. In Bakterien, Mitochondrien und Chloroplasten sind Angehörige der YidC/Oxa/Alb3-Proteinfamilie an Insertion und am dreidimensionalem Zusammenbau von Membranproteinen beteiligt.

In dieser Doktorarbeit wird eine direkte Interaktion zwischen YidC aus *E. coli* und dem translatierenden wie auch dem nicht-translatierenden Ribosom gezeigt. Eine vergleichbare Interaktion konnte für Oxa1 aus *Saccharomyces cerevisiae* (einem YidC-Homologen) und Ribosomen aus *E. coli* bestätigt werden. Es werden die mit Hilfe von Kryo-Elektronenmikroskopie bestimmten Strukturen von am translatierenden Ribosom gebundenen *E. coli* YidC beziehungsweise *S. cerevisiae* Oxa1 vorgestellt. In den Kryo-Elektronendichten befinden sich Dimere von YidC beziehungsweise Oxa1 oberhalb des ribosomalen Tunnelausgangs. Zusätzlich zeigen Crosslink-Versuche, dass das Ribosom Dimerzustände von YidC und Oxa1 spezifisch stabilisiert. Für die Funktion wichtige und evolutionär konservierte Transmembranhelices von YidC und Oxa1 wurden mit Hilfe von Cystein-Crosslink-Versuchen in der Kontaktregion der beiden

Monomere im Dimer gefunden. Interessanterweise kontaktieren sowohl Oxa1-Dimere also auch YidC-Dimere das Ribosome nahe dem ribosomalen Protein L23 und den evolutionär konservierten rRNA-Helices 59 und 24, wie es bereits für das nicht-homologe SecYEG-Translokon beobachtet wurde. Aufgrund dieser Resultate schlagen wir vor, dass Dimere von YidC beziehungsweise Oxa1 Insertionsporen bilden und mit dem SecY-Monomer den generellen Aufbau einer Proteintranslokations-/ Insertionspore teilen.

1 INTRODUCTION

1.1 The Bacterial Ribosome

Ribosomes translate DNA encoded genetic information into protein sequences in all kingdoms of life. The prokaryotic ribosome is a ribonucleoprotein complex with a molecular weight of approximately 2.5 MDa and a sedimentation coefficient of 70 Svedberg (70S). It consists of a large and a small ribosomal subunit, that sediment at 50S and 30S, respectively. Several high resolution structures of bacterial and archeal subunits as well as of the whole ribosome have been solved (Ban, Nissen et al. 2000; Wimberly, Brodersen et al. 2000; Schuwirth, Borovinskaya et al. 2005; Selmer, Dunham et al. 2006).

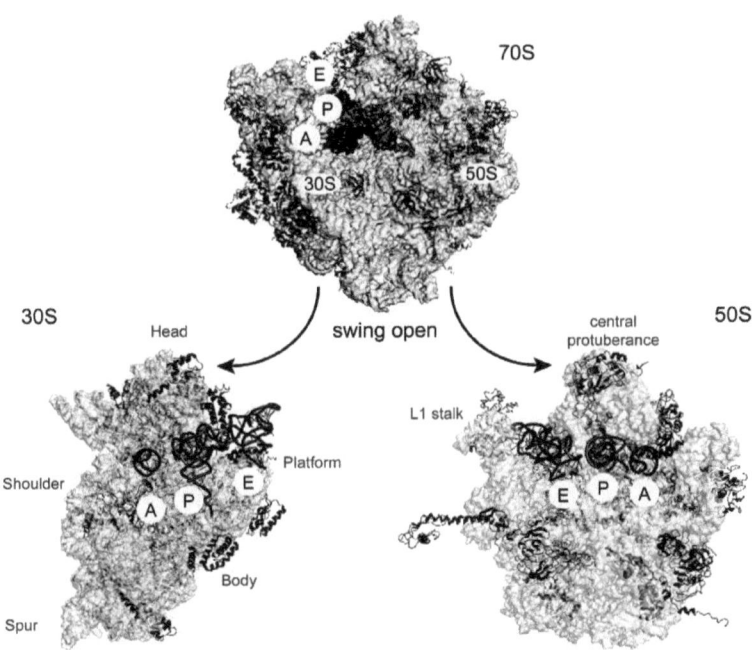

Figure 1: The prokaryotic ribosome. The *Thermos thermophilus* 70S ribosome comprises the 30S (PDB-entry 2J00) and the 50S (PDB-entry 2J01) subunits. Ribosomal RNA is shown in light grey; ribosomal proteins are depicted in dark grey. Labels A, P and E mark the binding sites of the aminoacyl-tRNA (dark grey), the peptidyl-tRNA (medium grey) and the exit-site tRNA (dark grey).

The peptidyl transferase center (PTC) of the ribosome, which is responsible for the formation of peptide bonds, is located within the large ribosomal subunit. The 30S subunit harbors the decoding center, where the messenger RNA (mRNA) codons interact with the complementary anticodons of the amino acid bearing transfer RNAs (tRNAs). The 70S ribosome comprises three different binding sites for tRNAs: the A-site accommodating the aminoacyl-tRNAs, the P-site, where the tRNA bearing the nascent chain is bound, and the E-site, serving as an exit site for de-acetylated tRNAs (Figure 1). The process of translation can be depicted as a three-stage process consisting of initiation, elongation and termination:

The translational process is initiated by formation of the 30S-initiation complex: The mRNA pairs via the Shine-Dalgarno sequence upstream of the start codon with a complementary sequence of the 16S rRNA in the 30S subunit (Shine and Dalgarno 1974). Initiation factors IF1 and IF3 prevent early rejoining of the 50S and stabilize the binding of the initiator tRNA (tRNAfMet) which is delivered by IF2 to the 30S subunit (Antoun, Pavlov et al. 2006). Promoted by IF2, the subunits join and thereby complete the 70S-initiation complex.

During elongation, an aminoacylated tRNA with an anticodon complementary to the codon on the mRNA is delivered to the ribosomal A-site by elongation factor Tu (EF-Tu). Upon correct codon-anticodon interaction, GTP is hydrolyzed on EF-Tu, which thereby releases the aminoacyl tRNA to relax into the PTC (Valle, Zavialov et al. 2003). The peptidyl transferase active site is formed exclusively by the 23S RNA of the large ribosomal subunit, with participation of the 3' terminal nucleotides of the tRNAs, which means that the ribosome actually is a ribozyme (Ban, Nissen et al. 2000; Nissen, Hansen et al. 2000). The nascent peptide chain is then linked to the tRNA in the A-Site, which is relocated to the P-site, whereas the anticodon loop of the deacetylated tRNA is moved to the E-site due to conformational changes induced on the ribosome by elongation factor G (Moazed and Noller 1989). The now empty A-site permits accommodation of a new aminoacyl tRNA, the first step of the next elongation cycle. The nascent chain emerges from the exit of the polypeptide exit tunnel when it has reached a length of about 35 amino acids (Hardesty and Kramer 2001).

Elongation is terminated when a translation stop codon on the mRNA reaches the A-site, which is recognized by release factor RF1 or RF2 (Youngman, McDonald et al. 2008). Binding of one of these release factors to the ribosome coincides with the release of the polypeptide chain from the P-site tRNA, whereupon the ribosomal subunits separate assisted by ribosomal recycling factors and EF-G (Hirashima and Kaji 1973).

1.2 Insertion of Nascent Membrane Proteins into the Membrane

A large proportion of proteins need to be incorporated into specific cellular membranes to fulfill their function; in *E. coli* these are about 20-30% of the total number (Luirink, von Heijne et al. 2005). Membrane proteins play an important role in diverse cellular processes, such as energy conversion, cellular transport, signal transduction and cell division. Like all other proteins, membrane proteins are synthesized by the ribosome. Membrane proteins usually have a high content of hydrophobic amino acids and thus risk aggregation in aqueous environments like the cytosol. Therefore, most membrane proteins are co-translationally inserted into the membrane, thereby avoiding contact of the nascent protein chain with an unfavorable cytosolic environment.

1.2.1 The SRP-SecYEG insertion pathway

Most nascent membrane proteins (IMPs) are recognized by the ubiquitous signal recognition particle (SRP) (Keenan, Freymann et al. 2001). The SRP binds to the hydrophobic targeting sequences in the nascent protein when they start emerging from the polypeptide exit tunnel of the ribosome (Figure 2). SRP subsequently targets its bound substrate to SecYEG (in bacteria) or Sec61 (in eukaryotes), which form protein-conducting channels (Brundage, Hendrick et al. 1990; Rapoport 2007) in the membrane. The pore-forming subunit of SecYEG, SecY (corresponding to Sec61α/p in eukaryotes), forms two lobes wherein the two parts could open laterally like a clam to release the substrate transmembrane helices into the surrounding bilayer (Clemons, Menetret et al. 2004; Van den Berg, Clemons et al. 2004; Bostina, Mohsin et al. 2005). During translocation, the polypeptide chain to be transferred passes through the center of the clam shaped protein (Cannon, Or et al. 2005; Becker, Bhushan et al. 2009). When inactive, the protein channel is sealed by a short "plug" helix, thereby maintaining the diffusion barrier created by the plasma membrane. In order to open the channel, the plug helix moves from its central position to a peripheral binding site (Robson, Carr et al. 2009).

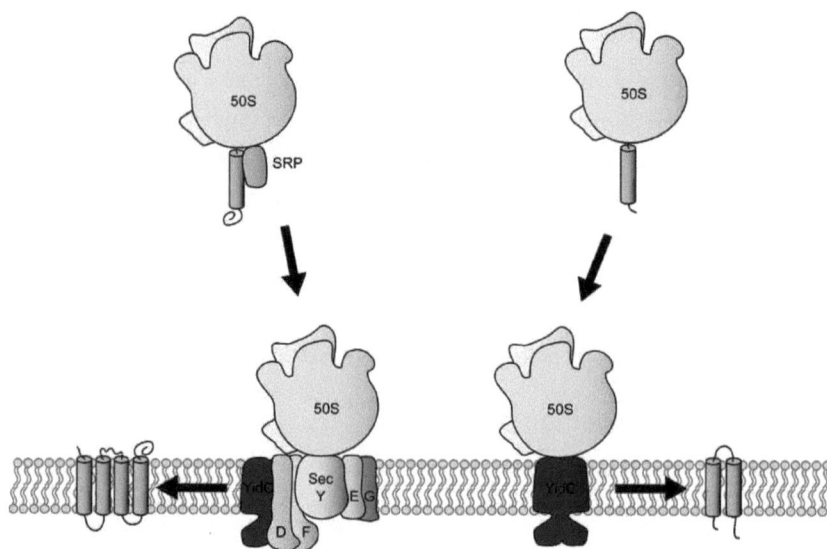

Figure 2: Insertion of membrane proteins in bacteria. The SRP-SecYEG pathway (or "Sec-dependent" YidC pathway) is shown on the left: SRP binds to the signal sequence of a nascent protein and delivers it to the SecYEG protein conducting channel. YidC is associated with the SecYEG translocon via SecDFYajC (YajC not shown) and likely assists in the release of the substrate into the surrounding lipid bilayer. On the right, the YidC-only (or "Sec-independent") pathway is shown where the ribosome docks on YidC, which then inserts the nascent protein helices into the membrane.

1.2.2 The role of the YidC/Oxa1/Alb3 protein family in membrane insertion

YidC and homologues thereof occur in all kingdoms of life, and in all studied cases are involved in membrane protein biogenesis (Luirink, Samuelsson et al. 2001). The energy transducing membranes of bacteria, mitochondria and chloroplasts have a high demand for the insertion and assembly of respiratory complexes and ATP synthases. This requires the function of a specific membrane protein 'insertase' from the YidC/Oxa1/Alb3 family (Bonnefoy, Chalvet et al. 1994; Samuelson, Chen et al. 2000; Scotti, Urbanus et al. 2000; Luirink, Samuelsson et al. 2001). In *E. coli*, YidC is essential (Samuelson, Chen et al. 2000). It works in conjunction with the canonical SRP-SecYEG pathway and also as a Sec-independent insertase (Figure 2). This latter function is similar to that of Oxa1 in mitochondria, which lack the components of the Sec system (Glick and Von Heijne 1996). Oxa1 therefore represents the main insertion pathway into the inner mitochondrial membrane; likewise for proteins that have been synthesized in the mitochondrial matrix or have been imported into the matrix from the cytosol.

1.2.3 Function of YidC in Sec-dependent membrane insertion

YidC has been found to be involved in membrane protein folding, assembly and quality control (Beck, Eisner et al. 2001; Nagamori, Smirnova et al. 2004; van Bloois, Dekker et al. 2008). In the Sec-dependent pathway, it interacts with transmembrane helices (TMHs) of nascent membrane proteins after their release from the SecYEG translocon (Urbanus, Scotti et al. 2001; van der Laan, Houben et al. 2001), thereby preventing their aggregation and possibly accelerating the insertion process. Thus, YidC has been proposed to function as a membrane chaperone (Nagamori, Smirnova et al. 2004). Membrane protein insertion probably occurs co-translationally since YidC was found to be associated with nascent Sec-dependent proteins (Samuelson, Chen et al. 2000; van der Laan, Houben et al. 2001; Froderberg, Houben et al. 2004). The interaction of YidC with the SecYEG translocon is thought to be mediated via the accessory complex SecDFYajC. YidC can form a stable heterotetramer with SecD, SecF and YajC, and probably interacts with the SecDF subunits, while SecF and YajC likely link the complex to SecYEG to form the holotranslocon (Nouwen and Driessen 2002). In the membrane, YidC is much more abundant than SecYEG (Driessen 1994; Urbanus, Froderberg et al. 2002). Therefore, it likely exists both as a complex with SecYEG-SecDFYajC and in an unbound form.

1.2.4 The Sec-independent YidC pathway of membrane protein insertion

In the Sec-independent pathway, YidC acts alone, and is responsible for the insertion of such diverse proteins as the F_0 subunit c of the ATP synthase (F_0c), subunit II of the cytochrome o oxidase (CyoA) and the mechanosensitive channel MscL (van der Laan, Urbanus et al. 2003; van der Laan, Bechtluft et al. 2004; Facey, Neugebauer et al. 2007) although in the latter example, YidC may only be involved in a late stage of insertion. In a recent study, cysteine accessibility assays suggested that depletion of SecYEG and surprisingly, also the depletion of YidC, did not significantly affect membrane insertion of MscL. Instead, YidC was found to be necessary for oligomerization of the MscL protein (Pop, Soprova et al. 2009). Therefore the authors proposed that in the case of MscL, YidC functions as a membrane chaperone.

YidC alone is also sufficient for the insertion of several small phage proteins that were previously thought to insert spontaneously into the membrane (Samuelson, Chen et al. 2000; Serek, Bauer-Manz et al. 2004). Recently, it has been suggested that YidC changes the tertiary structure of its membrane-spanning domain upon binding of the Pf3 coat protein (Winterfeld, Imhof et al. 2009). Substrates of the Sec-independent YidC pathway are typically rather small and often consist of two TMHs connected via a short hairpin (Kiefer and Kuhn 2007). To date, it has not been firmly established if the Sec-independent insertion route is exclusively co-translational and whether or not SRP is involved in Sec-independent substrate targeting (Kiefer and Kuhn 2007).

1.3 Structural Information on YidC

Members of the YidC/Oxa1/Alb3 family possess a conserved core of 5 TMHs (Figure 3 a, b), which define the insertase function (Jiang, Chen et al. 2003). The importance of the YidC/Oxa1/Alb3 proteins is reflected by their remarkable functional complementarity, which spans large evolutionary distances: Alb3, Cox18 and Oxa1 can complement the function of YidC, Alb4 can replace Oxa1 and YidC can substitute for Cox18 (Jiang, Yi et al. 2002; van Bloois, Nagamori et al. 2005; van Bloois, Koningstein et al. 2007). Gram-negative YidC proteins feature an N-terminal extension to the core insertase consisting of an additional TMH and a large periplasmic domain (Kiefer and Kuhn 2007).

The structural information on YidC is limited to a high-resolution structure of the soluble periplasmic domain (Oliver and Paetzel 2008; Ravaud, Stjepanovic et al. 2008) and a 10 Å projection map of the full length membrane bound dimer (Lotz, Haase et al. 2008) (Figure 3 c, d). In the crystal structure, the periplasmic domain adopts a β-supersandwich fold with an alpha-helical and, likely, flexible linker region at the C-terminal end. The N-terminal part of the domain seems to be flexibly linked to the non-essential YidC TMH 1 (Oliver and Paetzel 2008). In the crystal structure, the periplasmic domain is monomeric suggesting that YidC dimerization determinants reside in the transmembrane (TM) region of the protein. In the YidC projection structure a region of lower density was observed at the dimer interface and proposed to be part of the insertion pore (Lotz, Haase et al. 2008). Interestingly, YidC TMH 1 and TMH 3 were shown to contact TMHs of several YidC substrates and thus could form part of a potential pore or substrate binding region (Yuan, Phillips et al. 2007; Klenner, Yuan et al. 2008; Yu, Koningstein et al. 2008).

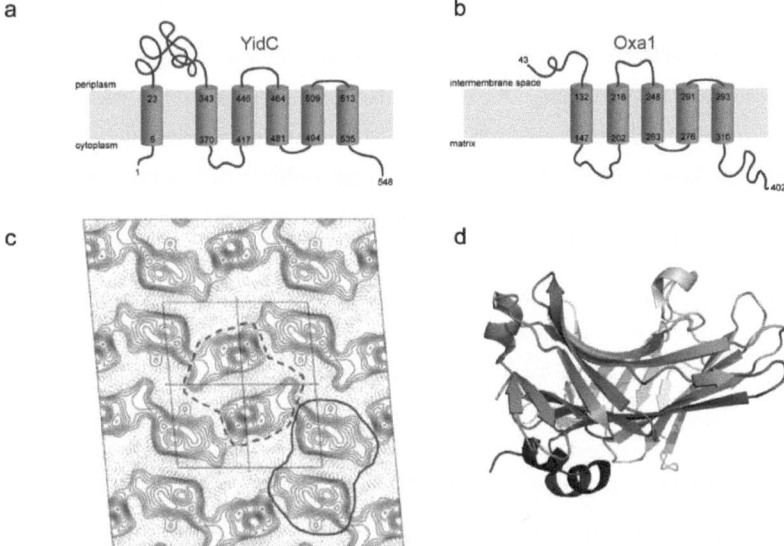

Figure 3: Structural information on YidC and Oxa1. (a) Topology overview for Escherichia coli YidC. **(b)** Topology overview for Saccharomyces cerevisiae Oxa1. **(c)** Projection structure from Lotz et al., showing a YidC dimer in the 2D crystal plane (Lotz, Haase et al. 2008). **(d)** Crystal structure from Ravaud et al., showing the periplasmic domain of E. coli YidC (Ravaud, Stjepanovic et al. 2008).

1.4 The Mitochondrial Oxa1 Protein

Mitochondria possess neither SRP nor SecYEG (Glick and Von Heijne 1996) and therefore no pathway homologous to the canonical SRP-SecYEG pathway that is responsible for the insertion of most membrane proteins in bacteria. Yet, mitochondria retain a need for a classical, bacterial-like insertion of the mitochondrially encoded proteins, all of them small TM proteins. Therefore, Oxa proteins must serve this function in mitochondria, in addition to facilitating the insertion of those membrane proteins that are imported into the matrix from the cytoplasm, like the F_0c homolog Atp9 (Jia, Dienhart et al. 2007). The Oxa1 protein contains a C-terminal extension that is required for ribosome binding (Jia, Dienhart et al. 2003). Furthermore, Oxa1 has been crosslinked to Mrp20, the mitochondrial homolog of the *E. coli* ribosomal protein L23, associating it with a co-translational mechanism of membrane protein insertion (Szyrach, Ott et al. 2003).

1.5 Aim of this Project

In contrast to the well-studied SRP-SecYEG pathway of membrane protein insertion, the YidC pathway was discovered more recently. As of yet, the insertion process has been studied biochemically for several model substrates of YidC, but the structural information on the insertase remained limited to a high resolution structure of the non-essential periplasmic domain and a medium resolution projection structure of the full length protein from a 2D crystal. In spite of the implication that YidC functions co-translationally, an interaction of YidC with the ribosome has not been shown before.

Among the important issues that still need to be solved are the following: Identification of the anchoring regions on both the ribosome and the insertase; determination of substrate binding regions of the YidC protein, which could elucidate the substrate specificity and possible targeting functions of YidC; the existence and nature of a translocation pore and lateral exit gate for the substrate helices; the oligomeric state of active YidC; the conformational changes between inactive and active states of YidC; maintenance of the diffusion barrier ("sealing") of the membrane during inactivity of the insertase; comparison with the Sec-translocon to work out if one "universal" or rather diverse kinds of insertion mechanisms exist. A cryo-EM reconstruction of the RNC-YidC complex would contribute to the general understanding we have of the YidC insertion system. Unfortunately, no atomic model of YidC or its homologues is available to date. In the ideal case, such a model of the protein could be fitted into the EM density in order to interpret detailed molecular interactions and possibly also conformational changes of the involved molecules.

During this study we focused on several issues: 1) The interaction between YidC and the ribosome. Binding experiments were designed in order to verify such an association and to obtain a complex stable enough for electron microscopic studies. 2) The binding regions on YidC. Possible ribosome binding regions include the positively charged C-terminus and the cytosolic loops of YidC. Here, the intended strategy was to delete short segments of the YidC protein and assess ribosome binding of the mutant. 3) A 3D cryo-EM reconstruction of the RNC-YidC complex. An RNC-YidC map would reveal contact areas between the binding partners and the shape of the YidC insertase. Further, comparison with the SecYEG-ribosome structure would enable us to compare the insertion mechanisms used by the YidC and Sec pathways. 4) The stoichiometry. In order to analyze the stoichiometry of the YidC-ribosome complex, crosslinking studies and native PAGE experiments were envisaged. Furthermore, the EM reconstruction should help in solving the stoichiometry issue. 5) A putative complex between Oxa1 and the ribosome. The mitochondrial YidC homologue Oxa1 lacks the periplasmic domain of YidC. An RNC-Oxa1 EM density therefore could be used to determine which regions of the YidC protein are represented in a RNC-YidC EM map. Oxa1 studies should also reveal a possible conservation of the insertion mechanism. This thesis describes the biophysical experiments and cryo-EM structure determination procedures employed to resolve the aforementioned issues.

2 MATERIALS

2.2 Chemicals

Compound	Purity	Company
2-mercaptoethanol	microselect	Fluka
2-propanol (isopropanol)	p.a.	Merck
Acetone tech.	technical	ETH
Acrylamide solution 40% 37.5:1	-	AppliChem
ADA	ultra	Fluka
Agar granulated	-	Difco
Agarose low EEO	-	AppliChem
Ammonium persulfate (APS)	mol.biol.grade	AppliChem
Ampicillin sodium salt	-	AppliChem
ATP	-	AppliChem
bacto®tryptone	-	Difco
bacto®yeast extract	technical	Difco
Boric acid	mol.biol.grade	AxonLab
Bromphenolblue sodium salt	90 %	AppliChem
Chloroform	-	Merck
Copper phenanthroline	-	Gift from I.C.
Coomassie brilliant blue	-	AppliChem
Coomassie G-250	-	Fluka
Cymal-6	anagrade	Anatrace
D(+)-Sucrose	for microbiol.	AppliChem
Dipotassium hydrogen phosphate anhydrous	p.a.	Merck
dNTPs, 10 mM	-	Fermentas
DTT	99.5 %	AppliChem
EDTA disodium salt dehydrate	mol.biol.grade	AppliChem
Ethanol absolute	-	ETH
Ethanol pure	99.8 %	Fluka
Ethidium bromide solution 1%	-	AppliChem
Glucose, anhydrous	for microbiol.	AppliChem
Glycerol	98 %	Synopharm
Glycine	99 %	Fluka
HEPES	99.5 %	Sigma
Hydrochloric acid (fuming)	puriss.	Fluka
Imidazole	99.5 %	Fluka
IPTG	99 %	AppliChem
Kanamycin sulfate	-	Fluka
Magnesium acetate tetrahydrate	ultrapure	AppliChem
Magnesium chloride hexahydrate	99%	Fluka
Methanol	technical	ETH

Compound	Purity	Company
n-Decyl-β-D-maltopyranoside	lyophilized	Glycon
n-Dodecyl-β-D-maltopyranoside	lyophilized	Glycon
Nickel sulfate hexahydrate	p.a.	Riedel-deHaën
Phenol	-	AppliChem
PMSF	BioChemica	AppliChem
Potassium chloride	p.a.	Merck
Potassium dihydrogen phospate	p.a.	Merck
Potassium hydroxide	ultra	Fluka
RNAse off	-	AppliChem
SDS	99%	Fluka
Sodium chloride	p.a.	Merck
Sodium dihydrogen phosphate hydrate	p.a.	Merck
Sodium hydroxide	98%	Fluka
TEMED	-	AppliChem
Trizma Base, Tris	99.9%	Sigma
Tween 20	-	AppliChem
X-gal	-	AppliChem

2.2 Enzymes

Restriction enzyme	Recognition site	Company
EcoRV	GAT/ATC	NEB
HindIII	A/AGCTT	Fermentas
NdeI	C/CATGG	Fermentas
PstI	CTGCA/G	NEB
XhoI	C/TCGAG	NEB

Other enzymes, antibodies and inhibitors	Company
T4 DNA ligase	Fermentas
Calf intestinal phosphatase	NEB
DNaseI	Roche
Picomaxx Polymerase	Stratagene
Pwo Polymerase	Roche
RiboLock™ RNase inhibitor	Pharmacia

2.3 DNA and Protein Molecular Weight Markers

100bp DNA ladder (NEB)
Range: The 100bp DNA ladder yields the following 12 discrete fragments:
1517, 1200, 1000, 900, 800, 700, 600, 500/517, 400, 300, 200, 100 bp
Concentration 0.5 mg DNA/ml

Gene Ruler™ 100bp DNA Ladder Plus (Fermentas)
Range: The 100bp DNA ladder Plus yields the following 14 discrete fragments:
3000, 2000, 1500, 1200, 1031, 900, 800, 700, 600, 500, 400, 300, 200, 100 bp
Concentration: 0.5 mg DNA/ml

Prestained Protein Molecular Weight Marker (Fermentas)

Protein	approx. MW	source
β-galactosidase	120 000	E.coli
Bovine Serum albumin	86 200	bovine plasma
Ovalbumin	47 000	chicken egg white
Carbonic anhydrase	34 000	porcine muscle
β-lactoglobulin	26 000	bovine milk
Lysozyme	20 000	chicken egg white

Novex® Sharp Protein Standard (Invitrogen)
260, 160, 110, 80, 60, 50, 20, 15, 10, 3.5 kDa

The patterns generated by the markers are shown in Figure 4.

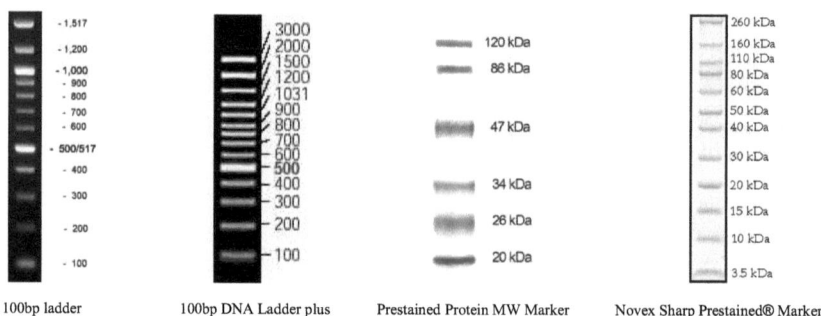

100bp ladder   100bp DNA Ladder plus   Prestained Protein MW Marker   Novex Sharp Prestained® Marker

Figure 4: Patterns of DNA and protein molecular weight markers used in this thesis

2.4 Buffers and Solutions

2.4.1 Antibiotics and inducers

Ampicillin (100 mg/ml)
1 g ampicillin sodium salt,
Millipore® water to 10 ml,
sterilize using a 0.2 µm sterile filter,
store aliquots at –20 °C

Kanamycin (100 mg/ml)
1 g kanamycin sulfate,
Millipore® water to 10 ml,
sterilize using a 0.2 µm sterile filter,
store aliquots at –20 °C

IPTG (100 mM)
0.282 g IPTG,
Millipore® water to 10 ml,
sterilize using a 0.2 µm sterile filter,
store aliquots at –20 °C
use 1 ml/l culture

2.4.2 Electrophoresis buffers

10x TBE
108 g Trizma base,
55 g boric acid,
40 ml 0.5 M EDTA pH 8.0,
ddH$_2$O to 1 l

6x DNA loading buffer
5 g sucrose ,
12 mg bromphenolblue sodium salt,
1x TE buffer to 10 ml

10x SDS-PAGE electrode buffer
30 g Trizma base,
144 g glycine,
10 g SDS,
ddH$_2$O to 800 ml,
adjust pH to 8.1 to 8.3 with HCl,
ddH$_2$O to 1 l

4x SDS-PAGE upper gel buffer
15.15g Trizma base,
1g SDS,
ddH$_2$O to 200 ml,
adjust pH to 6.8 with HCl,
ddH$_2$O to 250 ml

4x SDS-PAGE lower gel buffer
15.15 g Trizma base,
1 g SDS,
ddH$_2$O to 200 ml,
adjust pH to 6.8 with HCl,
ddH$_2$O to 250 ml

6x SDS sample loading buffer
7 ml 4x SDS upper gel buffer,
3 ml glycerole,
1 g SDS,
0.6 ml 2-mercaptoethanol,
1.2 mg bromphenolblue sodium salt

20x NuPAGE MES running buffer
1M MES,
1M Tris base,
69.3 mM SDS,
20.5 mM EDTA free acid

5x NuPAGE sample loading buffer
0.25 M Tris pH 8.5,
5 % w/v SDS,
50 % v/v glycerol,
0.51 mM EDTA,
bromphenolblue sodium salt,
10 mM DTT or 2- mercaptoethanol

20x NativePAGE running buffer
209.2 g BisTris,
179.2 g Tricine,
ddH$_2$O to 1 l

4x NativePAGE Sample buffer
0.418 g BisTris,
0.107 ml 6N HCl,
4 g glycerol,
0.117 g NaCl,
0.4 mg Ponceau S

20x NativePAGE cathode additive
1 g Coomassie® G-250,
ddH$_2$O to 250 ml

1x Dark Blue cathode buffer
10 ml 20x NativePAGE running buffer,
10 ml 20x NativePAGE Cathode Additive,
ddH$_2$O to 200 ml

2.4.3 Buffers for protein purification

Buffer A
20 mM ADA pH 5.8,
200 mM NaCL,
0.2% w/v DM

Buffer B
50 mM Hepes-KOH pH 7.5,
100 mM KOAc,

Buffer C
20 mM ADA pH 6.4,
150 mM NaCl,
0.1 % DDM

Solution I
20 mM Hepes-KOH pH 7.5,
150 mM NaCl,
100 mM Mg(OAc)$_2$

Buffer M
50 mM Na-phosphate pH 8.0 at 4 °C,
300 mM NaCl,
10 % v/v glycerol,

Buffer P
20 mM Hepes-KOH pH 7.5,
6 mM Mg(OAc)$_2$,
150 mM NH$_4$OAc,
2 mM spermidine,
50 µM sperminde,
2 mM 2-mercaptoethanol

Buffer R
20 mM Hepes-KOH pH 7.5,
200 mM NaCl,
25 mM Mg(OAc)$_2$,

Buffer TS
20 mM Tris-Cl, pH 8.0,
300 mM NaCl

Buffer TSG
20 mM Tris-Cl, pH 8.0,
300 mM NaCl,
10 % v/v glycerol

2.4.4 Other buffers and solutions

TE buffer
10 mM Tris-Cl, pH 8.0,
1 mM EDTA,
sterilize using a 0.2 µm sterile filter

CTAB/NaCl solution
4.1 g NaCl
10g CTA-Br
ddH$_2$O to 100 ml

2.4.5 Microbial growth media

Microbial growth medium	Composition
LB, 1 l	10 g bacto®tryptone 5 g bacto®yeast extract 10 g NaCl sterilize by autoclaving at 121 °C 100 µg/ml Amp and/or 25 µg/ml Kan
LB plates, 1 l	10 g bacto®tryptone 5 g bacto®yeast extract 10 g NaCl 6.5 g Agar, granulated sterilize by autoclaving at 121 °C 100 µg/ml Amp and/or 25 µg/ml Kan
TB, 1 l	12 g bacto®tryptone 24 g bacto®yeast extract 4 ml glycerol, water to 900 ml after autoclaving, add 100ml of a sterile solution of 0.17M KH$_2$PO$_4$ and 0.72M K$_2$HPO$_4$ 100 µg/ml Amp and/or 25 µg/ml Kan
SOC, 1 l	20g bacto®tryptone 5g bacto®yeast extract 5g NaCl 0.625ml 4M KCl 2.5ml 1M MgCl$_2$ 10ml 1M Glucose sterilize by autoclaving at 121 °C 100 µg/ml Amp and/or 25 µg/ml Kan

2.4.6 E. coli strains

Strain	Genotype	Application
DH5α	F⁻ endA1 hsdR17(rk⁻mk⁺) supE44 thi-1 λ⁻ recA1 gyrA96 relA1 $\Phi 80\Delta lacAm15$	Cloning, plasmid propagation & amplification
BL21(DE3)	F⁻ ompT hsdS$_B$ ($r_B^- m_B^-$) gal dcm (DE3)	Protein expression

2.4.7 Commercial kits

Bio-Rad Laboratories
Bio Rad Protein Assay
Details: *http://www.bio-rad.com*

Peqlab Biotechnologie GmbH
E.Z.N.A® Plasmid Miniprep Kit I (Classic Line)
PeqGOLD miniprep KitI
Details: *http://www.peqlab.de*

Invitrogen
TOPO® Cloning kit
NuPAGE 8-12% gradient gels
NativePAGE 3-12% native gels
Details: *http://www.invitrogen.com*

Qiagen
MinElute Gel Extraction Kit
Qiaquick Plasmid Midi Kit
Details: *http://www.qiagen.com*

Stratagene
Picomaxx® High Fidelity PCR System
Details: *http://www.stratagene.com*

2.4.8 Materials for cryo-EM and structure calculation

Balzer
High vacuum coating system (carbon gun)

DeLano Scientific LLC
PyMol^{TM} 0.99rc6
Details: *http://pymol.sourceforge.net/* and (DeLano WL 2002).

ETH Technical Services
Cryo-EM Grid Plunger
Custom built

FEI Company
FEI Tecnai F20 transmission electron microscope
Details: *http://www.fei.com/*

Image Science
Imagic-5 software package
Details: *http://www.imagescience.de*

Kodak
Electron Image Film SO-163 (8.3 x 11.9 cm)
Professional D-19 Film Developer (powder)
Details: *http://www.kodak.com*

Nikon
Nikon Super Coolscan 9000 ED scanner
Details: *http://nikonimaging.com/global/products/scanner/index.htm*

Quantifoil
Quantifoil® Grid R 2/1
Details: *http://www.quantifoil.com*

3 METHODS

3.1 DNA Methods

3.1.1 Preparation of genomic DNA from *E. coli*

Phenol/chloroform extraction followed by isopropanol precipitation was used to isolate and purify genomic DNA from *E. coli*. All centrifugation steps were carried out in an Eppendorf F45-30-11 rotor. 1.5 ml of a BL21(DE3) o/n culture were pelleted at 6'000 rpm for 4 min before resuspending the pellet in 567 μl TE buffer on ice. After addition of 30 μl of 10% (w/v) SDS solution and 3 μl of 20 mg/ml proteinase K, the mixture was incubated at for 1h at 37 °C. 100 μl of 5 M NaCl solution were added and the solutions were mixed thoroughly. The tube content was mixed with 80 μl CTAB/NaCl solution (0.7M NaCl, 10% CTAB) by inverting several times. After incubation at 65 °C for 10 min, an equal volume of chloroform was added and the sample was centrifuged at 14'000 rpm for 10 min.

In a fresh tube, the supernatant was mixed with an equal volume of phenol. The phases were separated by centrifugation (14'000 rpm, 5 min, 4 °C), the supernatant transferred into a fresh tube and completed with an equal volume of chloroform. 0.6 volumes of isopropanol were added to the extracted supernatant and the precipitate was pelleted (16'000 x g, 10 min, 4 °C). After washing the pellet with 1 ml of 70 % ethanol and pelleting at 14'000 rpm for 5 min at 4 °C, the supernatant was removed, the pellet air-dried and then dissolved in 100 μl TE buffer.

3.1.2 Preparation of plasmid DNA

Plasmid DNA was purified from *E. coli* DH5α cells using commercial DNA extraction kits. The peqGOLD miniprep KitI and the Qiagen plasmid Midi kit were used for the purification of small and medium amounts of plasmid DNA, respectively. Both kits rely on a combination of alkaline lysis and strong, but reversible binding of plasmid DNA to silica membranes under certain conditions. The protocols were carried out according to the manufacturers' manuals.

3.1.3 Agarose gel electrophoresis

Agarose gels for the analysis and purification of DNA were prepared by dissolving 1.0 -1.2 % (w/v) agarose in 60 ml of TBE buffer by heating in a microwave. One drop of a 1 % ethidium bromide solution was added to the dissolved agarose and the mixture was poured into a gel casting form. Solidified gels were run at a constant voltage of 90 mA in TBE buffer.

3.1.4 Polymerase chain reaction

Polymerase chain reaction (PCR) was used to amplify DNA fragments for cloning. 50-200 ng of template DNA were combined with 0.5 µM of the forward and backward primers, 0.2 mM of dNTP mix, 1 x PicoMaxx reaction buffer (Stratagene), and 2.5 U of PicoMaxx polymerase (Stratagene) in a final volume of 50 µL. Smaller reactions of 25 µl were used to optimize annealing temperatures.

An exemplary thermocycler program is listed below

3 min @ 94 °C

25-30 cycles of:
15 sec @ 94 °C
15 sec @ (lowest T_{melt} of used primers – 5 K)
1 min @ 72 °C (duration depending on template length)

10 min @ 72 °C
Cooling @ 4 °C

Agarose gel electrophoresis was used to purify the PCR products prior to ligation experiments.

3.1.5 Restriction digests

For restriction digests of DNA on an analytical scale 2 µl of Miniprep DNA were incubated with 10 U of restriction enzyme in the supplemented reaction buffer (Fermentas) in a total volume of 10 µl. The sample was incubated at 37 °C for 1 h and the cleavage products analyzed on an agarose gel. For restriction digests on a preparative scale 4 to 8 µg of DNA were incubated in a total volume of 50 µl at 37 °C for 3 h or over night. In some cases, plasmid DNA was dephosphorylated with calf intestinal phosphatase at 37 °C for 1 h to prevent self-ligation.

Agarose gel electrophoresis was used to confirm the size of digest products and the DNA of interest was extracted from the gel with commercially available gel extraction kits (Qiagen MinElute and QIAquick). The protocols were carried out according to the manufacturers' manuals.

3.1.6 TOPO® cloning (Invitrogen)

The Zero Blunt® TOPO® PCR cloning kit was used for blunt end subcloning of PCR products into the high-copy TOPO® vector. Thus, DNA fragments could easily be propagated and amplified in bacterial cells for sequencing and further cloning experiments.

In a typical experiment, 1 µl of TOPO® vector was incubated with 3 µl of purified PCR product and 1 µl of diluted salt solution at room temperature for 20 min in a total volume of 6 µl. 1-6 µl of the sample was then used for electroporation into *E. coli* DH5α and cells were plated onto LB agar. The agar contained 100 µg/ml of ampicillin; 4 µl of a 1 M IPTG solution and 40 µl of a 20 mg/ml X-gal stock solution were plated onto it directly before use.

3.1.7 Ligation of DNA fragments

For ligations, vector and DNA insert were combined in ratios between 1:5 and 1:10. 0.5-1 µg of restriction digested plasmid DNA was incubated with the respective amount of digested insert DNA in a total reaction volume of 10 µl, containing 1mM of ATP, 1 µl T4 ligase (NEB) in T4 ligase buffer (NEB). Ligation mixtures were incubated at room temperature for 1-4 h or at 16 °C over night.

3.1.8 Transformation of electro-competent *E. coli* cells

Electroporation was used to transform bacterial cells with plasmid DNA. Therefore, 50 µl of electro-competent *E. coli* cells were thawed on ice, mixed gently with up to 2 µl low-salt plasmid DNA or 2 µl of ligation mix. The actual electroporation step was carried out in a pre-cooled electroporation cuvette. 600 µl of SOC medium at room-temperature were added to the transformed cells. The cell suspension was incubated at 37 °C for 1 h and plated on selective LB agar plates.

3.1.9 DNA sequencing

Samples for DNA sequencing were sent to Synergene Biotech GmbH or Microsynth GmbH.

3.1.10 Expression of YidC and Oxa1 constructs

The YidC gene was amplified from genomic DNA of *E. coli* whereas an Oxa1 library plasmid was generously provided by M. Peter to serve as a PCR template. The gene encoding wild-type *E. coli* YidC (1644 bp) was cloned with a C-terminal hexa-histidine (His6-) tag into a pET expression Vector (pET24a_YidC$_{His6}$, Novagen). N-terminally His6-tagged plasmids pProEx_$_{His6}$YidC and pProEx_$_{His6}$YidCΔc were obtained by cloning the wild-type YidC gene or fragment bp 1 - 1605 of the wild-type gene, respectively, into the pProExHtb B vector (Invitrogen).

The C-terminally His6- and Myc-tagged plasmid (pBAD/Myc-His_YidC$_{Myc\_His6}$) was produced by cloning wild-type YidC (1644bp) into pBAD/Myc-His (Invitrogen); this experiment was done by I.C.

pET24a_Oxa1$_{His6}$ (Novagen) was obtained by cloning bp 127 - 1210 of the wild-type *S. cerevisiae* Oxa1 gene, which corresponds to the amino acid sequence of the mature Oxa1 protein after cleavage of the mitochondrial targeting sequence, into the pET24a expression vector. Expression conditions were similar for all pET24a constructs. Plasmids were transformed into *E. coli* strain BL21 (DE3) by electroporation. Cells were grown in TB medium at 37 °C, and induced with 100 µM IPTG at an OD$_{600}$ of 1.6. The induced culture was grown for another 12 – 15 h at 25 °C and harvested by centrifugation (5'000 x g, 10min).

E. coli C43 cells harboring the pBAD construct were grown by shaking at 37 °C in 2 × YT broth containing 100 µg/ml ampicillin. Over-expression was induced with 0.2% arabinose once the cells had reached an exponential phase of growth and the cells were harvested by centrifugation after 3 h (5'000 x g, 10min). The cell pellets were flash frozen in liquid nitrogen and stored at -80 °C.

3.1.11 Membrane preparation of YidC and Oxa1 constructs

To prepare bacterial cell membranes, cell pellets were thawed on ice, resuspended in buffer M containing 20 mM MgCl$_2$ and ruptured by two passes through a French Press. The lysate was cleared by centrifugation (10 min at 5'000 x g) and the resulting supernatant was centrifuged for 1 h at 100'000 x g to pellet membranes. The membrane pellet was homogenized in buffer M in a glass douncer.

3.1.12 Purification of YidC constructs

Bacterial membranes were extracted with 1% cymal® (Anatrace, Maumee, OH) in buffer TSG for 1.5 h at 4° C, then centrifuged for 15 min at 28'000 x g to remove insoluble material. The supernatant was loaded onto a Nickel affinity column (HisTrap™ FF Column 1 ml, GE Healthcare), washed with 30 ml TS buffer containing 0.2% w/v n-decyl β-D maltopyranoside (DM) and 30 mM imidazole, and eluted with 20 ml buffer TS supplemented with 0.2% w/v DM and 300 mM imidazole. The buffer was exchanged using a HiPrep™ 26/10 Desalting column (GE Healthcare) to Buffer A. The protein was flash-frozen and stored at -80° C at concentrations of 10-30 µM. The purification protocol was based on a procedure described previously (Lotz, Haase et al. 2008).

3.1.13 Purification of Oxa1

Bacterial membranes containing the Oxa1 construct were extracted with 1% w/v n-dodecyl β-D-maltoside (DDM) for 30 min at 4° C. After centrifugation for 15 min at 28'000 x g, the supernatant was loaded onto a 1 ml HisTrap™ FF Column, washed with 20 ml Buffer B and eluted with 20 ml buffer B supplemented with 300 mM imidazole. The buffer was exchanged for Buffer C using a HiPrep™ 26/10 Desalting column (GE Healthcare). The protein was flash-frozen and stored at -80 °C.

3.1.14 Preparation of membrane free cell extract

The protocol was based on the procedure described previously (Schaffitzel and Ban 2007). After combination of 70S ribosomes with the cytosolic fraction, a run-off translation reaction was conducted to ensure that no nascent peptide chains were bound to the ribosomes. 500 µl aliquots of cell extract were flash-frozen in liquid nitrogen and stored at -80 °C.

3.1.14 Preparation of *E. coli* 70S ribosomes

14 ml of membrane free cell extract (containing 60 mM KOAc) were thawed on ice, and puromycin (30 mM stock solution in buffer P) was added to a final concentration of 1mM. KOAc was supplemented in a final concentration of 0.5 M and the reaction mixture was incubated for 2h on ice. Centrifugation (4'000 rpm, 4 °C, 15 min, table top eppendorf centrifuge) was employed to sediment remaining bacterial cells and other particles.

2.4 ml of supernatant were loaded on each 10-50% sucrose gradient in solution I and centrifuged (23'000 rpm, 15 h, 4 °C, SW 32 rotor). The 70S band on the gradient (in the middle of the gradient tube) was harvested with a syringe and the ribosomes pelleted by ultracentrifugation (55'000 rpm, 3 h, 4 °C, TLA 55 rotor). Ribosomal pellets were dissolved in buffer R, flash-frozen in liquid nitrogen and stored at -80 °C.

3.1.15 Preparation of *E. coli* ribosome nascent chain complexes (RNCs)

RNCs were prepared as described previously (Schaffitzel and Ban 2007). After the final ultracentrifugation step, the pellet containing the RNCs was dissolved in buffer R and flash-frozen at a concentration of 1 µM. Before use, DM or DDM was added to a final concentration of 0.2% or 0.1% as required. RNCs for binding assays were purified by affinity chromatography only, omitting the sucrose gradient step (Schaffitzel and Ban 2007).

3.1.16 Sedimentation assays

100 – 420 nM of *E. coli* RNCs, 70S ribosomes or 30S ribosomal subunits (stored in buffer R at -80°C) were incubated with different ratios of YidC and Oxa1 in buffers A or C, respectively, supplemented with 25 mM MgOAc$_2$, for 1 h on ice. During a 3 h ultracentrifugation step (55'000 rpm, 4°C, TLA55 rotor) bound protein was pelleted together with ribosomes or ribosomal subunits. In the negative controls, ribosomes were replaced by an equal volume of buffer R supplemented with 0.2% DM or 0.1% DDM. The pellets or tube bottoms were washed once with buffer A or C, resuspended in 15 µl 1x SDS gel loading buffer and analyzed by SDS-PAGE and subsequent Coomassie staining. Positive controls were also applied and consisted of YidC, Oxa1 or ribosomes dissolved in 1x gel loading buffer.

3.1.17 Analysis of oligomeric state by native PAGE

15 µM stock solutions of YidC were used; the detergent concentration being either 0.2% or 1% DM. To increase the detergent concentration, 2 µl of the 0.2% DM YidC stock solution were incubated with 13 µl of 20 mM ADA pH 5.8, 150 mM NaCl supplemented with DM and 5 µl buffer R to simulate buffer conditions used for EM and the sedimentation assays. Samples with detergent concentrations between 1 x CMC and 15 x CMC were prepared (see lanes 1-6 in Figure 20). To decrease the detergent concentration in the second part of the experiment, 2 µl 1% DM YidC stock solution were diluted with 5 µl buffer R and 13 µl of 20 mM ADA pH 5.8, 150 mM NaCl with the appropriate concentration of detergent were added (see lanes 7-11 in Figure 20). All samples were incubated on ice for 1 h in order to reach equilibrium. Glycerol was added to a final concentration of 10% (v/v) to enable sample loading and the samples were subsequently analyzed on a NuPAGE native gel (Invitrogen; NativePAGE Novex Bis-Tris Gel System Manual) (Wittig, Braun et al. 2006).

3.1.18 PICUP crosslinking assays

Samples containing YidC or Oxa1 were crosslinked by the PICUP method (Fancy and Kodadek 1999). The buffer conditions were similar to those used in sedimentation assays; the total sample volume was 10 µl. Samples were prepared and preincubated for 30 min on ice. 4 mM ammonium persulfate and 0.2 mM Tris-bipyridylruthenium(II) were added to the mixture and the proteins were crosslinked by irradiation for 10 - 30 s with a 250 W slide projector lamp at a distance of 20 cm. The reaction was quenched by addition of 200 mM DTT and the samples were analyzed by 4-12% gradient SDS–PAGE (NuPAGE, Invitrogen) and Coomassie staining.

3.1.19 Disulfide crosslinking assays

Samples were prepared similarly to the protocol used for the PICUP crosslinking procedure. After the preincubation step, copper phenanthroline was added to a total concentration of 2 mM and the samples were incubated for 15 min at room temperature. The reaction was stopped by addition of the appropriate volume of 4x NuPAGE gel loading buffer and the reaction products were analyzed by non-reducing SDS-PAGE and subsequent Coomassie staining.

3.1.20 Sample preparation and data acquisition by cryo-electron microscopy

Sample preparation for cryo-EM was similar to that used for sedimentation assays. The concentration of *E. coli* RNCs was 100 nM and YidC or Oxa1 were added in an 8 - 12 fold excess. Samples were incubated for 30 min on ice before 3-5 µl were pipetted on a glow-discharged (25 mA, 30 seconds) Quantifoil R2/1 grid coated with continuous carbon foil. The sample was left to adsorb to the grid for 1-2 seconds before excess liquid was blotted away with wet filter paper. Plunging into liquid ethane (Dubochet, Adrian et al. 1988) vitrified the aqueous sample solution. The frozen grids were kept at liquid nitrogen temperature and imaged in a FEI F20 microscope (FEI, Hilsboro, OR) at a magnification of 50,000 x with an acceleration voltage of 200 kV. 104 Micrographs were recorded on Kodak SO-163 film (Eastman Kodak, Rochester, NY) under low dose conditions at 1.0 – 4.0 µm defocus. Negatives were developed during 12 minutes with Kodak D-19 developer solution. The images were scanned with a CCD scanner (Super Coolscan 9000 ED, Nikon Corporation, Tokyo, Japan) at a step size of 12.7 µm and coarsened to a final pixel size of 3.17 Å on the object scale.

3.1.21 Data processing and 3D reconstruction

In order to assess the quality of the recorded data, power spectra of the scanned negatives were calculated. Only negatives containing high frequency information up to 10 Å and without noticeable drift or axial astigmatism were chosen for semi-automated particle picking (Figure 5) with the autobox function of the program Boxer (Ludtke, Baldwin et al. 1999). Particles overlapping with others or picked from areas with poorly vitrified ice as well as particles showing very high contrast were deleted manually.

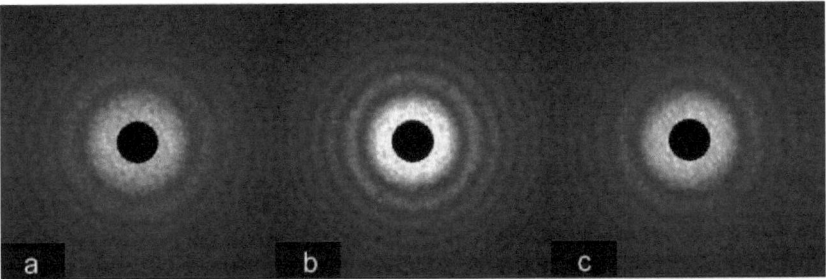

Figure 5: Exemplary power spectra of three negatives from the RNC-YidC data set. (a) and **(b)** were rated as good, with almost no drift or astigmatism. **(c)** was classified as poor, since information is lost in one direction due to drift.

The RNC-YidC data set consisted of a total of 45970 single particle images, whereas the RNC-Oxa1 data set comprised 19412 single particle images. Estimated theoretical power spectra were compared against power spectra computed from the picked images in order to find correct CTF parameters (Figure 6) (Sander, Golas et al. 2003). After CTF correction, the images were band-pass filtered and coarsened to a pixel size of 5.08 Å/pix for the first and to 3.17 Å/pix for the final rounds of calculation.

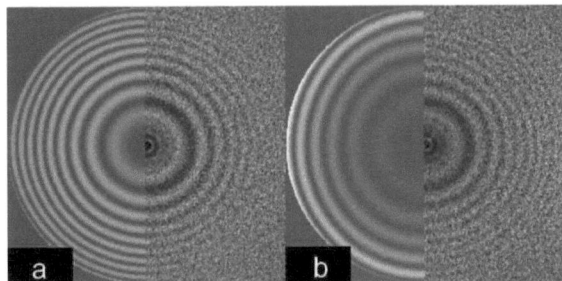

Figure 6: Comparison of the theoretically calculated amplitude spectrum (left half-images) with the amplitude spectrum computed from picked particle images (right half-images). (a) shows a good estimation of CTF parameters, since minima and maxima of both half-sides of the image match well. For **(b)**, the parameters were estimated poorly.

An RNC structure without bound factor was used as an initial reference map (Schaffitzel, Oswald et al. 2006). Using ***THRE-FOR***, this reference was projected to obtain reference projections which in the following step were needed to align particle images (in the following, Imagic-5 commands are written in bold italic letters while SPIDER commands are represented in normal bold letters).

In the initial rounds of calculation, the program IMAGIC (van Heel, Harauz et al. 1996) was used for all steps except for the alignment of particles that was done using a parallelized version of the **AP_SH** routine from the SPIDER package (Frank, Radermacher et al. 1996). For the experimental particle image, the in plane Euler rotation angle as well as translational shifts in X and Y directions were determined to enable alignment of the particle images with the reference projections.

Next, aligned particle images were grouped by correlation to the reference projections and groups of images were averaged with *SUM-ALI*. For the first three-dimensional reconstruction ("3d_ali") using *TRUE*, the known angles of the corresponding references were applied. *TRUE* is a 3D reconstruction program relying on filtered back-projection. Reprojection of 3d_ali yielded a set of projections with known angles ("anchor set") that would later be used for the determination of Euler angles of the class averages.

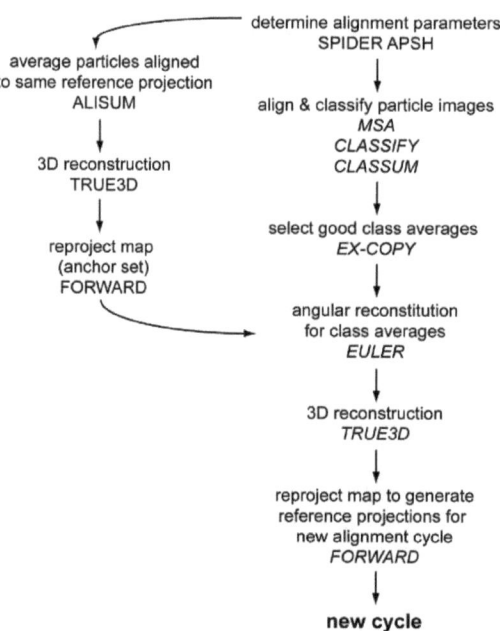

Figure 7: Exemplary IMAGIC calculation cycle.

Subsequently, particle images were subjected to multivariate statistical analysis by *MSA-RUN*, which calculates eigenimages and eigenvalues of a set of aligned particle images (a form of data compression to shorten computing time). ***MSA-CLASS*** then uses hierarchical ascendant classification to determine class-membership of the particles; the number of classes has to be specified by the user. Thereupon, images from each class were averaged by ***MSA-SUM*** to yield class averages showing reduced noise and thus improved signal in comparison to the individual single particle images that constitute the class members.

EULER uses the anchor set projection images to assign euler angles to the class averages by angular reconstitution. Knowing the Euler angles, class averages were used for filtered back projection and a new 3D reconstruction was generated ("3d_euler") by ***TRUE***. Finally, 3d_euler would become the reference map for a new cycle of alignment and 3D reconstruction.

After initial rounds of calculation using the IMAGIC routines, final rounds of refinement were done using the SPIDER software (Frank, Radermacher et al. 1996). For alignment, again **AP_SH** was used. Three dimensional reconstructions were done using the **BP_RP** and **BP_3F** commands.

The initial reconstructions of RNC-YidC and RNC-Oxa1 complexes showed only weak additional densities at the tunnel exit. This, together with analysis of the class averages, indicated that the RNCs were only partially occupied with YidC or Oxa1 respectively. To obtain homogenous particle subgroups supervised classification was employed (Figure 14) (Valle, Sengupta et al. 2002). In order to select for RNCs with a bound insertase, two reference maps were created: the RNC map with the weak additional density in the tunnel region served as "plus insertase" reference. For the "minus insertase" reference, the same RNC map was used, but the additional density at the tunnel was masked out. The two reference maps therefore resembled an RNC with bound insertase and an RNC without bound factor. Reference projections of both reference maps were then obtained as described above. Subsequently, two alignments were run so that every particle image was aligned to the "minus-insertase" reference projections as well as to the "plus-reference" projections. The data set could then be sorted by cross-correlation into particles that were more similar to the "minus-insertase" or to the "plus-insertase" reference. In the case of the RNC-YidC data set, finally 24395 particles were classified into the "plus" group, while 21575 particles ended up in the "minus" group. For the RNC-Oxa1 data set, the numbers were 8814 (plus) and 10598 (minus), respectively. Only the particles classified into the "plus" groups were used for further rounds of refinement. As a result, the densities for the insertases became much stronger. Since control maps calculated from the "minus" populations of particles showed no additional density in the ribosomal exit tunnel region, supervised classification succeeded in separating particles with and without bound insertases.

3.1.22 Generation of masks

Masks were used to cut off noise density not connected to the main density of the RNC-insertase complexes. In addition to this, the "minus" reference maps for supervised classification were generated by masking out the unwanted part of density, i.e. the additional density in tunnel region corresponding to the insertases. This was done by importing the respective density map into ImageJ (Abramoff, Magelhaes et al. 2004). The point tool was used to create an outline around the density to be "kept". Everything outside this outline would later be erased during the masking process. The file containing the point coordinates was converted into .plt format so that it could serve as input for the ***THRE-CONT*** program, where a binary mask was produced. In order to soften the mask, it was gauss-filtered by ***THRE-FIL***. The filtered mask could then finally be multiplied with the corresponding density map.

3.1.23 Extrapolation of the resolution value to the full data set

The resolution of the RNC-insertase complex depended on the number of particles included in the data set. For the calculation of FSC curves that are used to determine the resolution of EM maps, data sets are usually split in two halves. From each half-set, a density map is calculated and the two resulting maps are compared for different resolution shells in reciprocal space. This means that the resolution value obtained by this method describes the quality of a density map calculated with only half the number of particles. Since a correlation exists between the number of particles and the resolution (LeBarron, Grassucci et al. 2008), one can calculate resolution values for different numbers of particles and then extrapolate the value for the actual number of particles in the full data set.

4 RESULTS

4.1 Cloning of YidC and Oxa1 Constructs

The wild-type YidC gene was first cloned from *E. coli* genomic DNA into the pCR®II Blunt TOPO Vector (by N.D.). Further constructs usually comprised a His-tag for affinity purification and detection, and were based on the aforementioned pTOPO_YidC. All pET24 based constructs bear a cleavable N-terminal His-tag, while the YidC variants expressed in pProEx plasmids feature a C-terminal His-tag.

During the course of this project, the following YidC variants were cloned (Figure 8): pET24a_YidC$_{His6}$ encodes full length His-tagged YidC and was used for most binding assays and for EM grid preparation. pProEx_$_{His6}$YidC encoded full length YidC that was used as a control for binding tests involving N-terminally tagged truncation mutant $_{His6}$YidCΔc (encoded by pProEx_$_{His6}$YidCΔc), where the cytoplasmic C-terminus of YidC was deleted. Full length C-terminally His-tagged YidC$_{Myc\_His6}$ was provided by I.C. and was mainly used for cross-linking assays.

Mature *S. cerevisiae* Oxa1 lacks the N-terminal mitochondrial targeting sequence. pET24a_Oxa1$_{His6}$ encodes this mature form of Oxa1 that was used both for biochemical and EM experiments (Figure 8). A library plasmid was generously provided by M. Peter to serve as a PCR template.

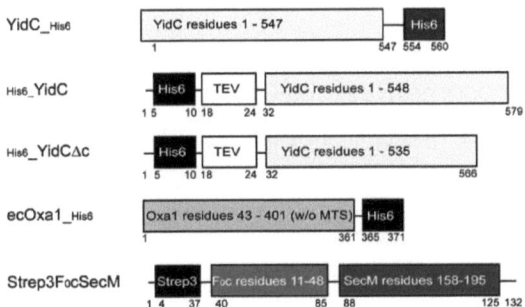

Figure 8: Construct overview showing YidC, Oxa1 and nascent chain expression constructs. The small numbers indicate the amino acid positions.

4.2 Cloning of F_0c Ribosome Nascent Chain Constructs

The E. coli F0c subunit of the F0F1 ATPase was used as a nascent chain bait in all experiments since F0c is a well-characterized substrate of YidC and can also be inserted into bacterial membranes by Oxa1 if Oxa1 is expressed in E. coli (van der Laan, Bechtluft et al. 2004; van Bloois, Nagamori et al. 2005). The Strep3F0cSecM nascent chain construct was designed such that only the first transmembrane helix of the F0c subunit (amino acids 1 to 79) is displayed on the ribosome, thereby effectively mimicking an insertion intermediate upon binding to the insertase (Figure 9). Strep3F0cSecM was cloned by C.S. into the pUC19 vector and includes an N-terminal triple Strep-tag for affinity purification.

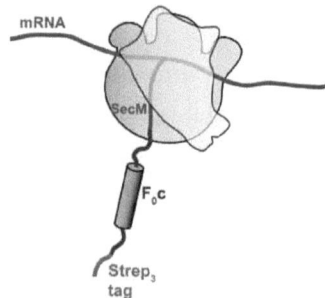

Figure 9: The F_0c ribosome nascent chain complex. The first transmembrane helix of the F_0c ATPase subunit is displayed on the ribosome. SecM stalling ensures a stable complex and the Strep-tag is used for affinity purification.

4.3 Preparation of Ribosome Nascent Chain Constructs

Ribosome nascent chain complexes (RNCs) for binding and cross-linking experiments as well as for EM grid preparation were generated by in vitro translation and purified by sucrose gradient centrifugation followed by affinity chromatography as described before (Schaffitzel and Ban 2007).

4.4 Purification of His-tagged YidC and Oxa1 Proteins

All pET24a and pProEx constructs were expressed in *E. coli* BL21 (DE3) cells at 37 °C. Cells were lysed by one passage through a French Press and debris was removed by centrifugation prior to membrane preparation. All purification steps were done at 4 °C. Thawed membranes were solubilized with decyl maltoside (YidC) or dodecyl maltoside (Oxa1) and loaded slowly onto HisTrap columns with a peristaltic pump. Elution fractions were concentrated to 1-2 ml volume and loaded onto a G25 Sephadex desalting column for buffer exchange (Figure 10). The protein eluted in a single peak, and desalting fractions were frozen separately in liquid nitrogen. Typically, 20 ml of *E. coli* cell membranes yielded about 3 mg of protein.

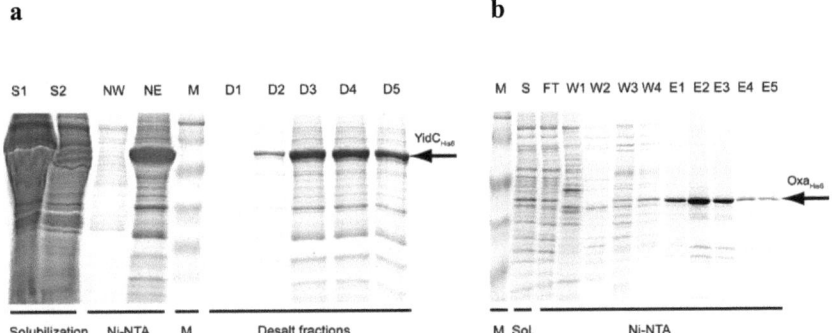

Figure 10: Purification of YidC$_{His6}$ and Oxa1$_{His6}$. (a) Representative example of a YidC purification: Ni-NTA chromatography followed by a desalting step. S1, solubilized membranes; S2, solubilized membranes after centrifugation, NW, Ni-NTA high salt wash; NE, Ni-NTA elution; M, Marker (120, 86, 47, 34, 26, 20 kDa); D1-D5, desalt fractions. (b) Representative example of an Oxa1 purification by Ni-NTA chromatography: M, Marker; S, solubilized membranes after centrifugation; FT, flow-through; W1-W4, wash fractions; E1 – E4, elution fractions. The Oxa1 purification was conducted by B.G.

4.5 The Cytoplasmic C-terminus of YidC Binds to the Ribosome

Particular substrates of the Sec-independent YidC pathway have been shown previously to insert co-translationally into the bacterial membrane (Chen, Samuelson et al. 2002; van der Laan, Bechtluft et al. 2004). To investigate this process we characterized the interaction of YidC with non-translating ribosomes and with ribosomes displaying the first transmembrane helix of the natural YidC substrate F_0c.

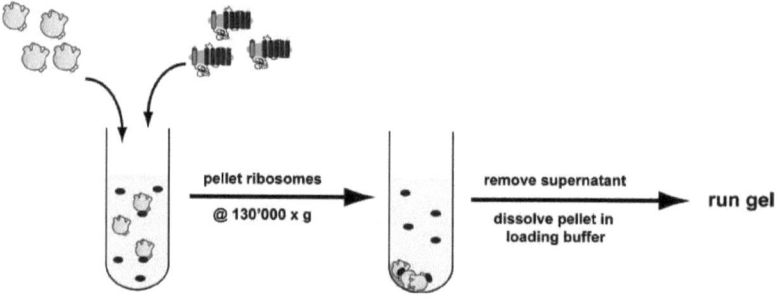

Figure 11: Ribosome sedimentation assay. Bound protein sediments along with ribosomes during ultracentrifugation. Pellets are then analyzed by SDS-PAGE.

For a typical ribosome sedimentation assay (Figure 11), purified *E. coli* 70S ribosomes or RNCs were incubated with detergent-solubilized, purified YidC. Subsequent analysis of co-sedimentation pellets after ultracentrifugation confirmed that *E. coli* YidC binds strongly to 70S ribosomes. Ribosome binding was further enhanced by the presence of the F_0c substrate nascent chain that additionally interacts with YidC (Figure 12a). Virtually no binding was detected between YidC and the small ribosomal subunit indicating that the large ribosomal subunit is required for a productive interaction (Figure 12b).

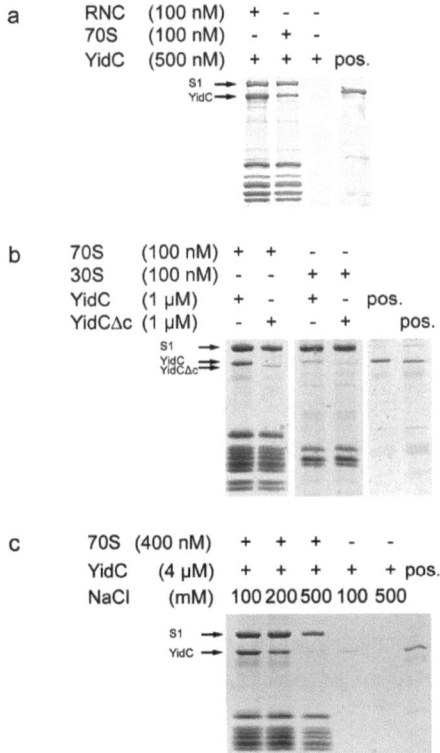

Figure 12: Binding of YidC to *E. coli* ribosomes and RNCs analyzed by ribosomal pelleting. The ribosomal pellet fraction obtained by ultracentrifugation was analyzed by SDS-PAGE and Coomassie staining. **(a)** YidC binds to ribosomes and RNCs. The positive control shows the YidC signal that would be expected for a 1:2 association. **(b)** The C-terminus of YidC is required for binding to the large ribosomal subunit. Positive controls show the YidC signals expected for a 1:1 association of YidC with the ribosome. **(c)** The ribosome-YidC interaction is salt-sensitive. The positive control shows the YidC signal expected for a 1:1 association.

The cytoplasmic loops of YidC contain a number of charged amino acids that could interact with the ribosome. In the case of Oxa1, the positively charged C-terminal domain has been shown to directly interact with the mitochondrial ribosome (Jia, Dienhart et al. 2003; Szyrach, Ott et al. 2003). In order to investigate whether the shorter, positively charged C-terminus of YidC (amino acids 536 to 548) could contribute to the ribosome binding in *E. coli*, the last 13 amino acids of YidC were deleted in the truncated YidCΔc protein. Ribosome binding was analyzed by co-sedimentation and SDS-PAGE. Deletion of the C-terminus reduced binding of YidC to ribosomes to background levels (Figure R12b). The interaction with YidC was also lost by increasing the ionic strength (Figure 12c) indicating that binding is mediated by ionic interactions. No significant interaction could be detected to the small ribosomal subunit (Figure 12b). Taken together, these data suggest that the charged C-terminus of YidC mediates important interactions with the large ribosomal subunit.

4.6 Oxa1 Interacts with the Bacterial Ribosome

Oxa1 has been shown to bind to mitochondrial ribosomes via its C-terminus (Jia, Dienhart et al. 2003; Szyrach, Ott et al. 2003) and it can restore growth in YidC depletion strains by taking over the Sec-independent function of YidC in *E. coli* (van Bloois, Nagamori et al. 2005). Therefore, this functional conservation was tested with respect to the *E. coli* ribosome. Indeed, we were able to establish a stable complex of the mature form of yeast mitochondrial Oxa1 with the bacterial ribosome. Similar to the results that were obtained in the YidC binding test, Oxa1 binds to ribosomes and RNCs (Figure 13). Although the difference is not as strong as in the YidC binding test, Oxa1 seems to prefer substrate ribosome nascent chain complexes over empty ribosomes.

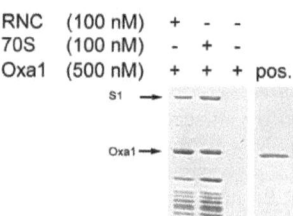

Figure 13: Binding of Oxa1 to ribosomes and RNCs analyzed by ribosomal pelleting. Oxa1 seems to bind preferentially to translating ribosomes. Experiment conducted by B. G.

4.7 EM Data Collection and Processing

In order to gain structural insight into the process of co-translational membrane protein insertion, we determined a 3D map of the complex between YidC and a SecM-stalled RNC by cryo-EM (Figure 15). Similar to the construct used for binding assays, the RNCs used for cryo – EM structure determination display only the first transmembrane helix of the F_0c subunit of the F_0F_1 – ATPase and thus mimic an insertion intermediate. In addition to the RNC-YidC complex, we also analyzed a RNC-Oxa1 complex by cryo-EM.

From film negatives rated sufficiently well (i.e. containing information to at least 10 Å, no strong drift or astigmatism), on average 1000 – 2000 particles per negative could be picked with the "autobox" function of the program boxer. In total, the RNC-YidC data set contained 45970 single particle images. The program IMAGIC was used for initial rounds of calculation based on a reference RNC structure without bound factor (Schaffitzel, Oswald et al. 2006). The reference RNC map was low pass filtered to ensure that the alignment was dominated by the lower frequency range representing the coarser features of the ribosome, which should be similar for both the reference and the RNC-YidC complex.

In the following, the program SPIDER was used for refinement of the initial 3D map and supervised classification was employed to separate subpopulations within the heterogeneous data set (Figure 14). For a typical supervised classification step, reference projections produced from two differing EM densities were correlated to individual particle images after alignment. The particle image was then assigned to one of the two subpopulations depending on how well it correlated to the two references.

For RNC-YidC, initial rounds of alignment and 3D reconstruction were done with the whole data set until the density was stable, i.e. did not change over several rounds of refinement. This initial EM map already showed some additional density in the exit tunnel region if compared to a ribosome without bound factor. Supervised classification was then used to discriminate between particles with and without YidC density at the exit tunnel region, the factor-less ribosome and the initial map showing weak YidC density serving as reference 3Ds. 53 % of all particles were assigned to the "plus YidC" subpopulation, and only these particles were used for the final rounds of refinement. Most likely, the "minus YidC" subpopulation contained not only ribosomes without bound YidC, but also ribosomes where YidC was not fully docked and therefore not resembling the "standard" conformation of the RNC-YidC complex. For RNC-Oxa1, the procedure was similar: Here, 45% of the 19412 particles were assigned to the "plus Oxa1" subgroup and used for the final rounds of refinement.

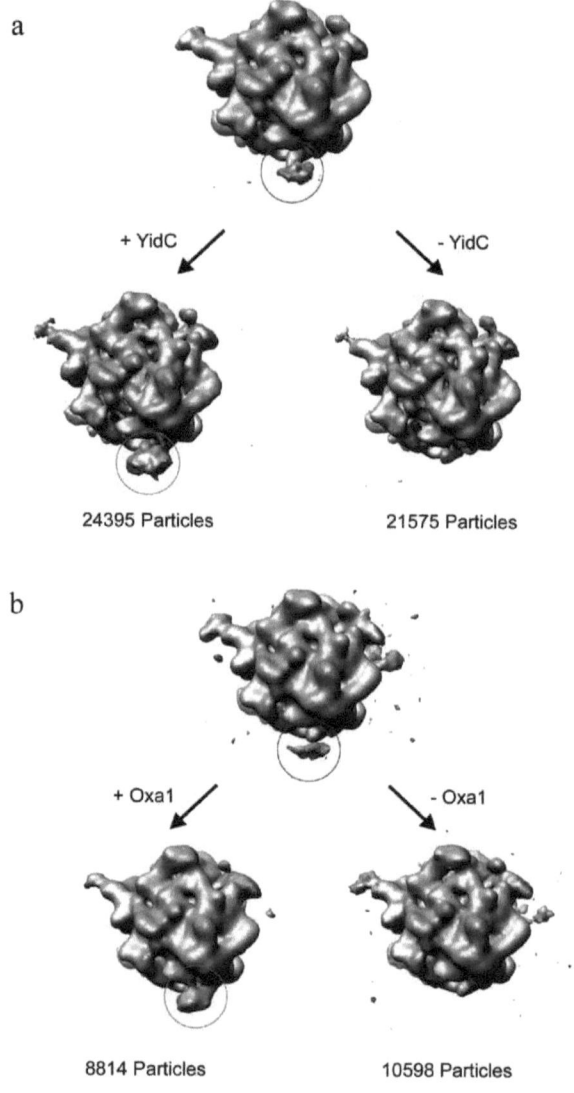

Figure 14: Supervised classification for RNC-YidC (a) and RNC-Oxa1 (b). The upper 3Ds show the maps resulting from initial calculations with the complete data set. The lower 3Ds show the refined maps calculated with the "plus factor" particle subgroups (left) and the "minus factor" particle subgroups (right).

4.8 3D Reconstruction of YidC Bound to the Translating Ribosome

The RNC-YidC cryo-EM map was refined to a resolution of 14.4 Å (Fourier shell correlation (FSC) 0.5 criterion); therefore the quality of this reconstruction is comparable to recent reconstructions of SecM-stalled ribosome nascent chain complexes (Figure 18) (Mitra, Schaffitzel et al. 2005; Schaffitzel, Oswald et al. 2006). The ribosomal P- and E- sites are occupied by transfer RNAs (tRNAs) whereas the occupancy of the A-site is low (Figure 15). Consequently, the growing polypeptide chain is predominantly attached to the peptidyl-tRNA in the P-site.

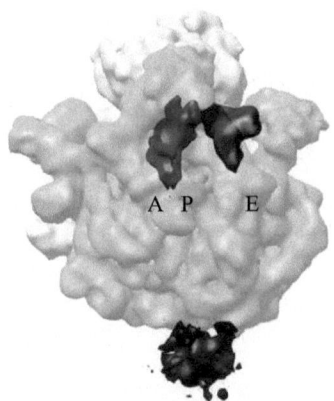

Figure 15: P-site and E-site are occupied by tRNAs in the Sec-M stalled RNC. View through the 50s subunit (translucent, foreground): P-site (left) and E-site (right) tRNAs are nicely visible, while there is no density visible for a tRNA in the A-site at the same threshold. Thus, the polypeptide chain is attached to the P-site tRNA in the Sec-M stalled RNC. YidC is shown in dark grey at the bottom.

The YidC density is located centered above the polypeptide tunnel exit and has a compact, slightly elongated bilobal shape with a diameter of about 70 Å (Figures 16b & 17). In both figures, the YidC density is shown at the same threshold level as the RNC thereby revealing the details of the interaction site. The two strongest connections of YidC to the exit tunnel region are visible in the L23/L29 region and near Helix 59 of the 23S rRNA (Fig. 16b).

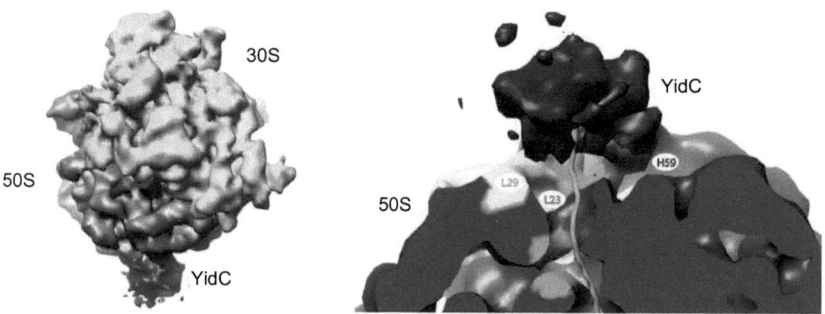

Figure 16: The 3D structure of YidC in complex with the translating ribosome. (a) Cryo-EM density of the *E. coli* RNC-YidC complex. YidC is bound at the tunnel exit of the. The small subunit is shown in light grey, the large subunit in medium grey and the YidC density in dark grey. (b) The ribosome was cut along the polypeptide tunnel to allow a direct view on the main contact sites of YidC. Ribosomal proteins located in this area are shown in shades of grey: L23 in light grey, L29 in medium grey, helix 59 of the 23S rRNA in medium to dark grey. The path of the modeled nascent polypeptide chain is shown as a light grey tube; it enters the YidC density at its center.

In Figure 17c, an outline of the YidC density depicted in Figure 17a is shown together with grey lines representing the membrane boundaries; the view is along the membrane plane. Since the dimension of the YidC insertase density agrees with the thickness of a lipid bilayer Figure 17c, the TM part of the YidC protein accounts for most of the EM map. As no density is visible above the membrane plane (the region corresponding to the periplasm in a cellular environment), this means that the probably flexibly linked, large periplasmic domain of YidC (Oliver and Paetzel 2008) is not represented in our RNC-YidC map. This is most likely due to the single particle 3D reconstruction process where flexible parts of the structure are averaged out.

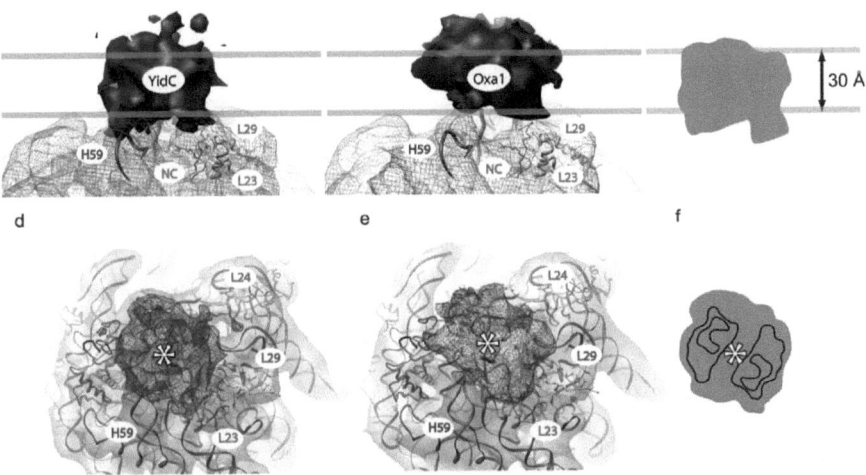

Figure 17: YidC and Oxa1 bound to the translating ribosome. Side view of YidC (**a**) and Oxa1 (**b**) (dark surfaces) on the bacterial ribosome (mesh). The horizontal lines indicate the membrane boundaries. Ribosomal proteins L23, L29 and L24 as well as helix 59 of the 23S rRNA are shown for orientation (compare with lower panel). (**c**) For comparison, the outline of the YidC density is shown together with the dimension of the plasma membrane. (**d, e**) Top view of YidC (**d**) and Oxa1 (**e**) along the ribosomal tunnel. (**f**) Superposition of the YidC projection structure (Lotz, Haase et al. 2008) onto an outline of the YidC density as shown in (**d**). The *E. coli* ribosome crystal structure (Schuwirth, Borovinskaya et al. 2005) was fitted into the ribosomal cryo-EM density. Ribosomal proteins L23, L29 and L24 as well as helix 59 of the 23S rRNA are shown in light grey. The star marks the exit point of the nascent protein chain (NC) on the ribosomal surface; its path is shown as a tube.

The volume enclosed by the density of YidC corresponds to a protein mass of two times the YidC TM portion. Since the YidC density is centered above the tunnel exit, the nascent protein chain, not visualized at this resolution, is likely to enter the YidC density in the middle (Figure 17d). Our reconstruction demonstrates a direct interaction between the ribosome, the nascent chain and the insertase in the Sec-independent YidC insertion pathway.

Three major contact sites were identified between the 50S ribosomal subunit and YidC (Figures 17a and 17d): Strong contacts are visible at helix 59 of the 23S ribosomal RNA (rRNA) and ribosomal proteins L23 and L29. The third contact site consists of two thinner connections that are formed at the tip of ribosomal protein L24 and at helix 24 of the 23S rRNA. In comparison to the *E. coli* 70S crystal structure (Schuwirth, Borovinskaya et al. 2005), helix 59 is displaced towards YidC by 6 Å

(Figure 17a). A similar conformational change has been observed in the RNC-SRP complex (9 Å displacement) (Schaffitzel, Oswald et al. 2006) and in the 70S-SecYEG complex (5 Å displacement) (Menetret, Schaletzky et al. 2007) suggesting some similarity in the mode of binding between all of these factors involved in co-translational protein targeting and translocation. The significance of this conformational change remains to be determined.

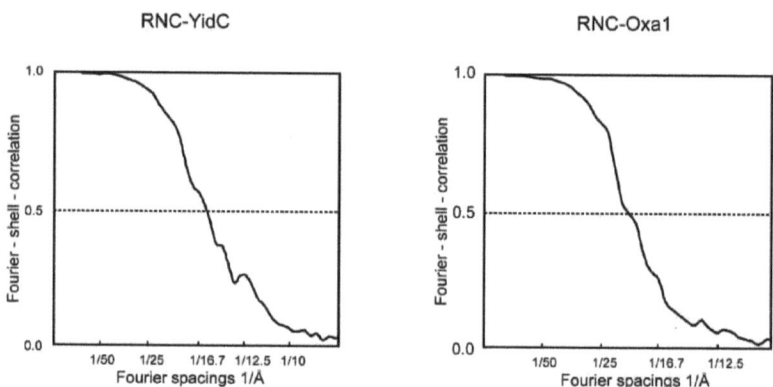

Figure 18: Diagram of the FSC function computed between two independent three-dimensional reconstructions. (left) Diagram for the RNC-YidC complex (continuous line). A set of 24396 particles was randomly split in two half sets to calculate the two reconstructions. The dotted line marks the FSC 0.5 threshold. (right) Respective Diagram of the FSC function for the RNC-Oxa1 complex. A set of 8814 particles was split in two half sets for the two independent reconstructions.

4.9 The RNC-Oxa1 Complex

To investigate whether in addition to their functional similarity in protein insertion, *E. coli* YidC and *S. cerevisiae* Oxa1 also share a common architecture and arrangement when in complex with the translating ribosome, we determined the 3D structure of Oxa1 bound to the *E. coli* RNC (Figure 19) to a resolution of 18.4 Å (Figure 18). For sample preparation, we again used *E. coli* RNCs displaying the F_0c subunit of the F_0F_1 ATP synthase, which is a homolog of the natural Oxa1 substrate Atp9 and can be inserted into the membrane by Oxa1 (Jia, Dienhart et al. 2007).

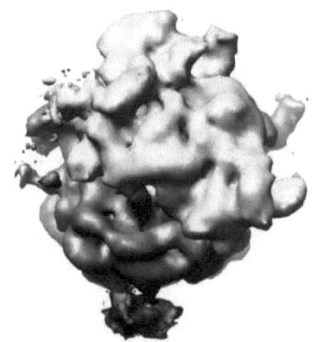

Figure 19: Cryo-EM Density of the RNC-Oxa1 Complex. Oxa1 is bound at the tunnel exit of the RNC, comparably to YidC. The small ribosomal subunit is shown in light grey, the large subunit in medium grey and the Oxa1 density in dark grey.

In the structure of the RNC-Oxa1 complex, the Oxa1 density is centered above the ribosomal tunnel exit. The size (diameter of ~70 Å) and shape of the Oxa1 density are similar to that of YidC (Figure 17a, b, d and e). The threshold level of the RNC-Oxa1 density was chosen at a level similar to that used for the RNC-YidC complex in order to enable a comparison between both structures. In the cryo-EM structure, Oxa1 is positioned similarly with respect to the ribosomal tunnel exit and oriented in the same way as YidC. The thickness of Oxa1, comparable to YidC also in this aspect, corresponds to that of a typical helical TM domain (Figure 17c). YidC and Oxa1 share a conserved core of 5 TM segments that are responsible for the insertase function. However, the size of the periplasmic part differs between both proteins: ~300 residues and ~90 residues for YidC and Oxa1, respectively. Therefore, the similar size of the two insertase densities confirms that the TM core dominates the observed density in both structures (RNC-YidC and RNC-Oxa1).

The contact sites of Oxa1 on the ribosome are almost identical to that of YidC. Again, a strong contact is visible at L23 and L29 (Figures 17c and e). This is consistent with the observation that Oxa1 can be crosslinked to the mitochondrial counterpart of *E. coli* L23, Mrp20 (Jia, Dienhart et al. 2003), probably by means of its ribosome binding C-terminus (Jia, Dienhart et al. 2003; Szyrach, Ott et al. 2003). Therefore, the contact at L23 is likely to be formed by the Oxa1 C-terminus. The second ribosomal contact site of Oxa1 is helix 59 of the 23S rRNA, which is displaced towards Oxa1 as described before for the RNC-YidC complex. The third connection between rRNA helix 24 and Oxa1 is rather weak, but can be observed at lower threshold values.

4.10 Dimers of YidC and Oxa1 Bind to the Ribosome

The dimensions of ribosome bound YidC match the structure of the membrane bound form of YidC, which is a dimer (Lotz, Haase et al. 2008) (Figure 17f). To further investigate the oligomeric state of YidC alone in solution and in complex with ribosomes, we performed Blue Native Gel Electrophoresis (BN PAGE) and chemical crosslinking experiments.

4.10.1 Analysis of the oligomeric state of YidC in solution by BN-PAGE

Purified YidC was incubated with increasing concentrations of decyl maltoside (DM) and then diluted again before being analyzed by BN PAGE (Figure 20). At the critical micellar concentration (CMC) of DM, a ladder of bands was visible, corresponding to monomer, dimer and higher oligomeric states. At higher detergent concentrations YidC migrated predominantly as a monomer with some dimers. When the detergent concentration was lowered again in the second part of the experiment, the monomer re-associated to dimers and higher oligomeric states. Thus, equilibrium exists between YidC monomers and dimers in detergent solution; the additional higher molecular weight forms are probably a result of aggregation due to loss of detergent micelles during dilution.

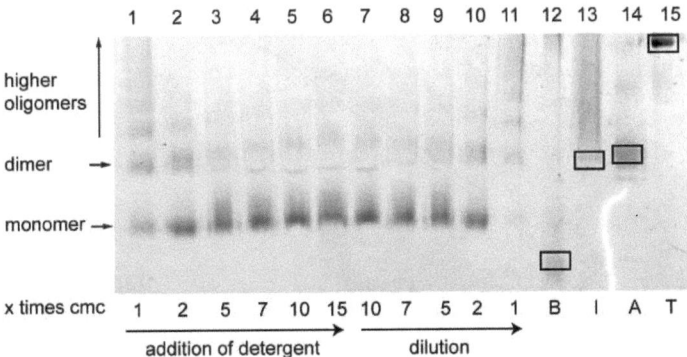

Figure 20: YidC exists in an equilibrium of monomers and dimers. Lanes 1-6: YidC stock solution (0.2%) was incubated with increasing concentrations of DM (corresponding to 1x CMC - 15x CMC). **Lanes 7-11**: YidC stock solution (1% DM) was diluted with buffer to decrease the detergent concentration. Samples were loaded on a native gel (Wittig, Braun et al. 2006). **Size standards, lanes 12-15:** BSA (B, ca 60 kDa), IgG (I, ca 150 kDa), Aldolase (A, ca 160 kDa) and Thyroglobulin (T, ca 700 kDa) were loaded as size references. The discrete bands in lanes 4-7 are likely detergent artifacts.

4.10.2 Analysis of the oligomeric state of YidC on the ribosome by Ru(bpy)$_3$ crosslinking

In order to assess the oligomeric state of YidC when bound to the translating ribosome, we used the PICUP crosslinking method. It utilizes the light-inducible crosslinking reagent tris-bipyridylruthenium(II) (Ru(bpy)$_3$) (Fancy and Kodadek 1999). Light activation of Ru(bpy)$_3$ together with ammonium persulfate (APS) triggers radical reactions (Fancy and Kodadek 1999) that induce fast and efficient covalent bond formation only between molecules that are in van der Waals contact. YidC with and without ribosomes and ribosomal subunits was incubated at a detergent concentration of 10x CMC before the PICUP reagents were added and the sample exposed to visible light generated by a slide projector lamp. Corresponding crosslinking experiments for Oxa1 were performed in parallel.

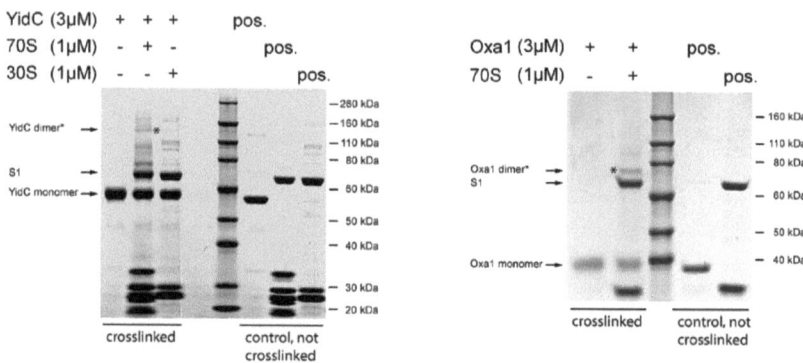

Figure 21: The ribosome stabilizes dimeric forms of YidC and Oxa1. [Ru(bpy)$_3$]$_2$-crosslinking of the RNC-YidC and RNC-Oxa1 complexes. Ribosomes were incubated with a 3-fold molar excess of YidC or Oxa1, APS and [Ru(bpy)$_3$]$_2$ prior to photo-activation. The crosslinked samples were analyzed by SDS-PAGE and Coomassie staining. The identity of the YidC and Oxa1 dimer bands (star) was confirmed by mass spectrometry. Control samples to the right of the marker bands were not crosslinked. **(a)** In the presence of ribosomes, a YidC dimer band appears, while in the 30S sample and the ribosome-free control, almost no dimer can be observed. **(b)** Similar to YidC, Oxa1 dimer formation could be observed in the presence of *E. coli* ribosomes.

Analysis of the crosslinked samples by SDS-PAGE showed that solubilized YidC and Oxa1 are primarily monomeric in the detergent concentrations used for EM and sedimentation experiments. In the presence of ribosomes, we observed an increased intensity of the dimer bands of YidC (Figure 21a) and of Oxa1 (Figure 21b). The relative intensities of the dimer bands obtained for YidC and Oxa1 can not be compared quantitatively since the two proteins require different sample buffers. In control experiments with the small 30S ribosomal subunit, the very weak intensity of the dimer band corresponds to the intensity observed for the insertase-only (i.e. ribosome-free) sample (Figure 21a). This suggests that the large ribosomal subunit stabilizes the dimer form of the YidC and Oxa1 insertases.

4.11 Cysteine Crosslinking of YidC and Oxa1 Dimers

In order to further characterize the dimerization interface of the conserved TM part of YidC and Oxa1, we assessed the vicinity of cysteines in opposing monomers by their ability to be oxidized and crosslinked to one another. YidC and Oxa1 each contain a single – in both cases conserved - cysteine, C423 in YidC and C141 in Oxa1 (Figure 22). Cysteine 423, located in TMH 3 of YidC, has been shown to crosslink to the nascent chain (van der Laan, Houben et al. 2001), indicating that it could be part of a potential insertion pore.

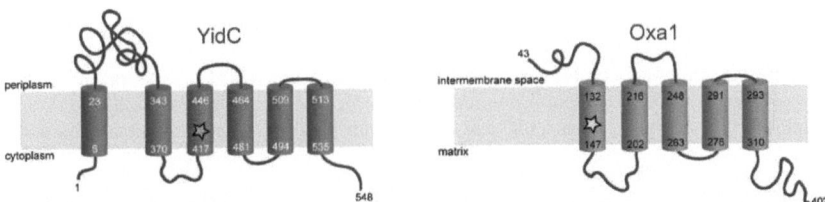

Figure 22: Positions of the single cysteines in YidC and Oxa1. In YidC, the single cysteine C423 is located in the third TMH of the protein, whereas the single cysteine of Oxa1 (C141) is located in the first TMH of the protein that is homologous to the second TMH of YidC. The cysteine positions are marked with stars.

Prior to performing the crosslinking reaction, purified YidC was incubated with or without ribosomes in order to allow the binding partners to associate. Copper phenanthroline is a hydrophobic oxidizing agent able to penetrate into membrane interiors and other hydrophobic environments. After incubation with this crosslinking reagent the reaction products were visualized by SDS-PAGE. For reactions containing only solubilized YidC, but no ribosomes, a weak dimer band could be observed (Figure 23a); if ribosomes were added to solubilized YidC, the fraction of molecules forming dimers increased considerably. This is in agreement with the $Ru(bpy)_3$ crosslinking experiments suggesting that the ribosome induces dimerization or stabilizes the dimeric form of YidC.

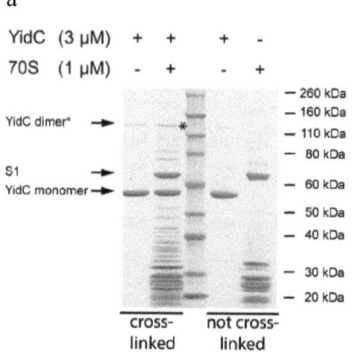 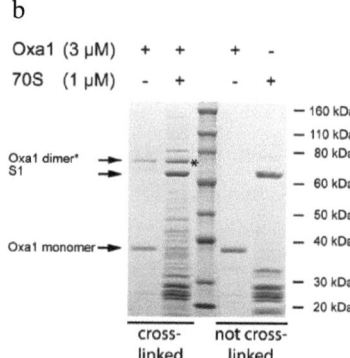

Figure 23. Cu-phenanthroline Oxidation of Cysteines in YidC and Oxa1. YidC **(a)** and Oxa1 **(b)** can be crosslinked as dimers by disulfide oxidation. Samples were analyzed by SDS-PAGE and Coomassie staining. The identity of the YidC and Oxa1 dimer bands (star) was confirmed by mass spectrometry.

Cysteines can only be crosslinked if the distance between the β-carbons is ~ 4.5 Å (Vinayagam, Pugalenthi et al. 2004) indicating that the two TMH 3 helices where cysteine 423 is located are very close to each other at the YidC dimer interface. Combining our cysteine crosslinking result with the finding that the nascent polypeptide chain contacts TMH 3 of YidC near C423 (Yu, Koningstein et al. 2008) suggests a positioning of the nascent chain in the center of the YidC dimer. The cysteine crosslinking experiment was repeated with Oxa1 (Figure 23b). Here, the single cysteine (C141) is located in TMH 1 (Figure 22), corresponding to TMH 2 in YidC. Alone, i.e. if no ribosomes were added, Oxa1 after oxidation was mostly in a monomeric form. In the presence of ribosomes, most Oxa1 is crosslinked as a dimer. Therefore, TMH 1 of Oxa1 also forms part of the dimer interface. We assume that the 3D arrangement of the conserved TMH core of YidC and Oxa1 is similar since YidC and Oxa1 are homologous – the TMH core being the most strongly conserved part of the protein - and in addition to that, are able to perform each other's function if expressed in the respective organism (van der Laan, Bechtluft et al. 2004; van Bloois, Nagamori et al. 2005; van Bloois, Koningstein et al. 2007). Our cysteine crosslink results imply that the two conserved TMHs 2 and 3 of YidC (corresponding to TMH 1 and 2 in Oxa1) form the central core of the dimer and are in close proximity to their counterparts in the second YidC molecule (Figure 24). This also suggests that the two YidC molecules dimerize in a head-to-tail organization, also indicated by the arrangement of YidC molecules in the membrane (Lotz, Haase et al. 2008).

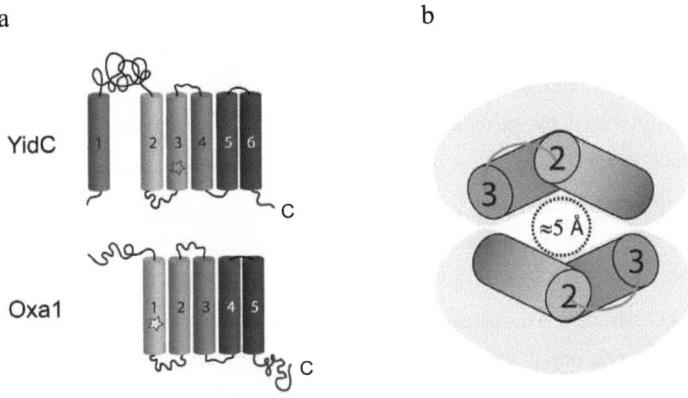

Figure 24: Model for TMH arrangement in the center of a YidC dimer. (a) Topology models of YidC and Oxa1. The conserved 5 TMH-core is shown together, and the positively charged C-termini starting at the end of the most C-terminal helix are marked. Homologous helices are shown in similar shades of grey. The position of the natural YidC cysteine C423 is shown as medium grey star. The natural Oxa1 cysteine C141 is marked with a light grey star; its corresponding residue in the YidC molecule is I368. **(bB)** Model of the arrangement of TMHs in the center of a YidC dimer. Assuming the 3D arrangement of the conserved TMH core is similar in YidC and Oxa1, both conserved helices TMH 2 and TMH 3 form the core of the dimer center and contact their counterparts in the second YidC molecule, thereby possibly contributing residues to an insertion pore. The distance between the C_β atoms of two crosslinked cysteines must be 5 Å or less. The blue kidney-shaped shadow symbolizes the rest of the TMH core.

5 DISCUSSION

This study reports the structures of YidC and Oxa1 bound at the polypeptide tunnel exit of translating bacterial ribosomes determined by cryo-EM. Our data show that the active form of the insertases is conserved and dimeric. The attachment sites on the ribosome are also consistent between YidC and Oxa1 and are shared by the non-homologous SecYEG translocon. This astonishing similarity shared by bacteria, mitochondria and presumably also chloroplast (Alb3) membrane insertion factors (Jiang, Yi et al. 2002), suggest a common architecture for protein insertion and translocation.

The mitochondrial and *E. coli* insertases share a conserved core of 5 TM segments (Figure 25); Oxa1 lacks TMH 1 and the large periplasmic domain of YidC, but has an extended highly positively charged C-terminus. The latter is important for insertion of nascent chains and makes a direct contact with mitochondrial ribosomes (Jia, Dienhart et al. 2003; Szyrach, Ott et al. 2003). Removal of the shorter, also positively charged C-terminus in YidC reduces the affinity of binding such that a stoichiometric RNC-YidC complex can no longer be observed in co-sedimentation experiments. However, a C-terminally truncated YidC retained activity *in vivo* indicating that other parts of the protein, as well as the substrate nascent polypeptide chain, provide weaker but sufficient interaction sites for ribosome binding (Jiang, Chen et al. 2003); all consistent with the multiple contacts observed in the structure. In addition, the SRP/FtsY system could contribute to targeting of YidC substrates to the insertase (Kiefer and Kuhn 2007). In mitochondria, in the absence of the SRP/FtsY system, the extension of the Oxa1 C-terminus might provide additional affinity to mitochondrial ribosomes. A recent study reported that chloroplast Alb3 uses its C-terminus for an interaction with chloroplast SRP43 and thus is involved in post-translational targeting of proteins to the chloroplast membrane (Falk, Ravaud et al.). In general, it seems that YidC/Oxa1/Alb3 proteins use their positively charged C-termini to engage in co-translational membrane targeting in bacteria, mitochondria and chloroplasts.

As the size and shape of the YidC and Oxa1 densities are similar, the density corresponds to the conserved TMH core. Thus, the large periplasmic domain of YidC that is probably flexibly connected to the TM part (Oliver and Paetzel 2008) was not resolved. This suggests that, due to its direct interactions with the ribosome the transmembrane domain is ordered while the peripheral features are somewhat flexible; a feature also apparent in the projection map of the membrane bound YidC (Lotz, Haase et al. 2008). Indeed, our solution structures and the projection map are compatible with each other and define a dimer formed by the TM portion of the molecule.

The active SecYEG complex is probably formed by membrane bound dimers (Bessonneau, Besson et al. 2002; Breyton, Haase et al. 2002). However, only one of them is active at any one time during post-translational translocation (Osborne and Rapoport 2007). In view of the mixed reports regarding the oligomeric state of SecY in translocon-ribosome complexes, (Mitra, Schaffitzel et al. 2005; Menetret, Schaletzky et al. 2007; Rapoport 2008), we decided to investigate the stoichiometry of the YidC and Oxa1 in complex with the ribosome using crosslinking assays in addition to the obtained structural data.

Photochemical and oxidative crosslinking experiments demonstrate that once again YidC and Oxa1 behave in the same way. Both experiments identify dimers that are stabilized by the large ribosomal subunit. Photochemical crosslinking using the PICUP method allowed us to detect the oligomeric state of the insertases under near-equilibrium conditions. The formation of the inter-molecular disulfide bonds between the naturally occurring cysteines in the YidC and Oxa1 proteins were particularly informative as they implicate specific TM segments at the dimer interface. Each protein contains a single natural cysteine, YidC in TMH 3 and Oxa1 in TMH 1 (equivalent to TMH 2 of YidC) (Figure 25).

Genetic studies indicate the importance of YidC TMH 3 and its close proximity to TMH 2 (Yuan, Phillips et al. 2007). YidC T362E (in TMH 2) was identified as an intragenic suppressor that overcame the cold sensitivity of the C423R mutation (in TMH 3), probably by forming a salt bridge (Figure 25). Therefore, TMHs 2 and 3 of YidC must be very close to each other and to their counterparts in the other ribosome-associated monomer. In this respect, it is particularly interesting that the nascent chain contacts TMH 3 (Yu, Koningstein et al. 2008). Thus, these four helices appear to form part of an insertion pore in YidC (Figure 25). The results also imply that the insertase dimer has a head-to-tail organization with TMH 2 and TMH 3 in the center, as already suggested by the 2D projection map. This is consistent with the position of the tunnel exit in the RNC-YidC complex where the nascent chain emerges beneath the center of the YidC dimer (Figure 17d and f).

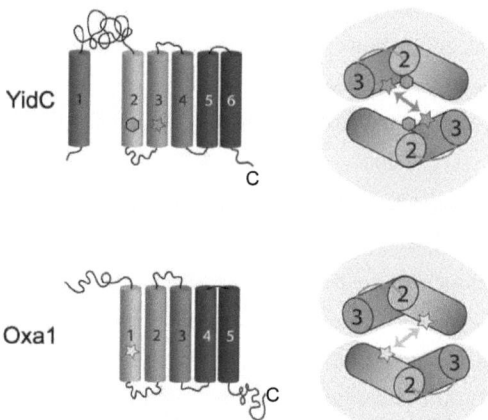

Figure 25: TMH 2 and 3 of YidC likely contribute to an Insertion pore. (left) Topology models of YidC (above) and Oxa1 (below). The conserved 5 TM-core is shown compacted, and the positively charged C-termini are marked. Homologous helices shown in similar shades of grey. The position of the natural YidC cysteine C423 is shown as an medium grey star. YidC T362 is marked with a hexagon. It is thought to be in close proximity to C423 (Yuan, Phillips et al. 2007). The natural Oxa1 cysteine C141 is marked with a light grey star, its corresponding residue in the YidC molecule is I368. **(right)** Arrangement of the TM helices in the center of a dimer. Assuming the 3D arrangement of the conserved TMH core is similar in YidC and Oxa1, both conserved helices TMH 2 and TMH 3 form the core of the dimer center and contact their counterparts in the second molecule, thereby possibly contributing residues to an insertion pore. The distance between the C_β atoms of two crosslinked cysteines must be 5 Å or less.

Recent data from Mathieu et al. have suggested additional functional links between the TMHs of Oxa1 (Mathieu, Bourens et al. 2010). The authors found several intragenic suppressors that compensated for point mutations otherwise leading to respiratory deficiency. In their study, they suggest functional links between TMH2 and TMH5, TMH4 and TMH5 as well as TMH4 and the loop between TMHs 2 and 3. In most cases, the complementation between a pair of mutant and compensating revertant suggest a spatial proximity of the corresponding positions, particularly if these two residues could form a salt bridge. However, the residues affected in Mathieu et al. are all nonpolar, and may form hydrophobic interactions. Thus, it is also possible that a point mutation interferes with the dimerization of Oxa1 while the revertant restores dimerization ability without the two positions actually being spatially close. Together with the results from Yuan et al. (Yuan, Phillips et al. 2007) who propose a salt bridge between TMHs 2&3 of YidC, and our findings that TMHs 1 and 2 of Oxa1 contact their counterparts in the second Oxa1 molecule of an Oxa1 dimer, it is possible to suggest a tentative helix arrangement for Oxa1 (Figure 26). In order to definitely resolve the issue of helix arrangement, a crystal structure of the protein would be critical.

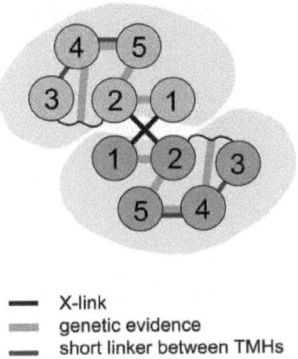

— X-link
═ genetic evidence
— short linker between TMHs

Figure 26: Tentative helix arrangement for Oxa1. The arrangement is based on the following information: Cysteine crosslinks (black lines) can be generated between TMHs 1 of an Oxa1 dimer and between TMHs 3 of a YidC dimer, corresponding to TMH 2 of Oxa1. Short linker sequences (medium grey lines) connect Oxa1 TMHs 3 and 4 as well as TMHs 4 and 5. Genetic evidence was found for functional links (light grey lines) between YidC TMHs 2 and 3 (corresponding to Oxa1 TMHs 1 and 2) and the following TMH pairs of Oxa1: TMHs 2 and 5 as well as TMHs 4 and 5. Furthermore, a functional link seems to exist between TMH4 and the loop between TMH2 and 3 which probably folds back into the membrane.

The cryo-EM map shows that the TM segments of both YidC and Oxa1 are anchored to the ribosome by three contact sites. The main connection stabilizing the interaction between YidC and the ribosome may involve the 13-residue short, positively charged C-terminus of YidC. As high-affinity binding is salt-sensitive, the association is most likely mediated by ionic interactions. The strongest connection in the L23/L29 region of the RNC-Oxa1 complex is possibly mediated by the Oxa1 C-terminus, as the biochemical data imply (Jia, Dienhart et al. 2003; Szyrach, Ott et al. 2003). An Oxa1-like C-terminal extension is also found in YidC of the bacteria *Rhodopirellula baltica* (Kiefer and Kuhn 2007) and *Streptococcus mutans* (Dong, Palmer et al. 2008). Therefore, it probably contacts areas that are conserved between bacterial and mitochondrial ribosomes, like L23.

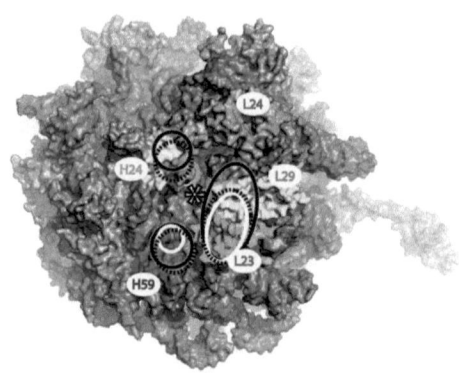

Figure 27: Comparison of YidC, Oxa1 and SecYEG bound to ribosomes. Contact areas of YidC, Oxa1 and SecYEG on the ribosome, view upon the 50S subunit with the tunnel opening in the center (star). The docking regions of the three insertases are highlighted as follows: YidC as dashed line, Oxa1 in white, SecYEG in black. Helix 59 (H59) of the 23S rRNA is a contact point for all three insertases, as well as the L23/L29 region. YidC and SecYEG have an additional strong anchor point at L24/H24, whereas Oxa1 shows only a weak connection in this area. Ribosomal proteins and RNA regions in contact with YidC, Oxa1 or SecYEG are marked. Helix 59 and helix 24 of the 23S rRNA are labeled as H59 and H24, respectively.

The three contact points of YidC and Oxa1 with the ribosome are the same ones used by the SecYEG translocon (Mitra, Schaffitzel et al. 2005; Menetret, Schaletzky et al. 2007) (Figure 27). All of them contact and perturb helix 59 of the 23S rRNA (Menetret, Schaletzky et al. 2007) and interact with a second site close to L23/L29. The latter region is a particularly frequently used docking site as it is involved in the interactions of several factors involved in co-translational folding, targeting and translocation (Schaffitzel, Oswald et al. 2006; Merz, Boehringer et al. 2008). A third connection for YidC and SecYEG is visible at L24 and helix 24 of the 23S rRNA, but is very weak in the RNC-Oxa1 structure.

Nascent chain substrates of the SecYEG translocon insert between two pseudo-symmetric halves of the SecY subunit (TMHs 1-5 and TMHs 6-10) (Van den Berg, Clemons et al. 2004). The two lobes may open like a clam, and thereby form a lateral gate for the passage of TM segments into the lipid bilayer. A superposition of the projection structure of YidC (Lotz, Haase et al. 2008) with the crystal structure of the archeal Sec complex (Van den Berg, Clemons et al. 2004) reveals the remarkably similar size of the SecYEβ complex and a YidC dimer; each consisting of 12 TMHs (Figure 28).

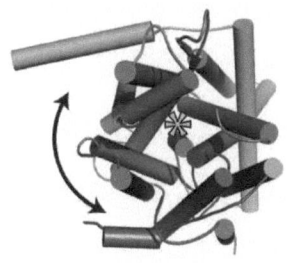 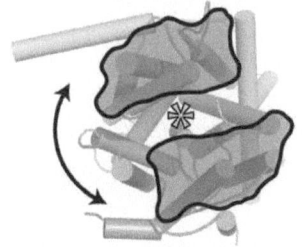

Figure 28: The YidC dimer shares similarities with the SecY monomer. Superposition of the YidC dimer projection structure outlines (Lotz, Haase et al. 2008) with the SecYEβ translocon crystal structure (Van den Berg, Clemons et al. 2004). The YidC dimer is similar in size to the SecYEβ translocon. The clam-like opening movement of SecYEβ is indicated by an arrow.

The two structures can be arranged such that the orientation of the respective 'half-channels' match. Therefore, we propose that the two YidC molecules in the dimer could function in a similar way to the two halves of SecY (Van den Berg, Clemons et al. 2004), with a channel in the center that opens by a separation of the component halves (Figure 29). The area of low density close to the interface in the YidC (Lotz, Haase et al. 2008) monomer may contribute to such a channel. In the Sec-independent pathway, YidC could directly bind its substrates and act analogous to SecYEG. Conversely, the results obtained for YidC suggest that although a dimer of SecYEG has been visualized bound to a translating ribosome (Mitra, Schaffitzel et al. 2005) it is possible that only one copy of SecYEG is active in co-translational protein translocation in a mechanism analogous to what was observed for post-translational translocation (Osborne and Rapoport 2007). Interestingly, newer ribosome-translocon reconstructions suggest that a single copy of the translocation complex is bound to the ribosome. Additional density around the translocon is attributed to the detergent micelle (Menetret, Schaletzky et al. 2007; Menetret, Hegde et al. 2008; Becker, Bhushan et al. 2009).

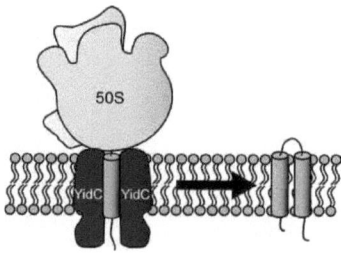

Figure 29: The YidC insertase functions as a dimer. Based on the crosslink results, the active YidC membrane insertase is a homodimer. It probably works similarly to a SecY monomer (pseudodimer), by creating a lateral opening to the membrane surroundings through which the substrate can leave the insertion pore.

In the Sec-dependent pathway, YidC most likely acts downstream of SecYEG (van der Laan, Houben et al. 2001). Based on our structures, the ribosome-associated SecYEG translocon would preclude the binding of YidC. In this case, the interaction between SecYEG and YidC is probably mediated by the SecD/F/YajC complex (Nouwen and Driessen 2002). A mechanism could be envisaged where substrate molecules are passed on laterally from SecYEG translocon to the interface of the YidC dimer.

Using a combination of cryo-EM and biochemical methods we show that dimers of the YidC-like protein insertases and the SecYEG translocon show a remarkable similarity in the architecture and the mechanism of ribosome binding. Since there is no evidence that SecYEG and the YidC/Oxa1/Alb3 protein family are related by evolution, their common mode of ribosome binding may be a consequence of their related function in co-translational insertion of nascent protein chains into a membrane. The results reported here provide the structural basis for additional biochemical and genetic experiments on this system and pave the way towards future understanding of YidC and Oxa1 mediated protein insertion at high resolution.

6 APPENDIX

6.1 DNA and Protein Sequences of YidC and Oxa1 Constructs

YidC$_{His6}$

DNA sequence

```
1     ATGGATTCGC AACGCAATCT TTTAGTCATC GCTTTGCTGT TCGTGTCTTT CATGATCTGG
61    CAAGCCTGGG AGCAGGATAA AAACCCGCAA CCTCAGGCCC AACAGACCAC GCAGACAACG
121   ACCACCGCAG CGGGTAGCGC CGCCGACCAG GGCGTACCGG CCAGTGGCCA GGGGAAACTG
181   ATCTCGGTTA AGACCGACGT GCTTGATCTG ACCATCAACA CCCGTGGTGG TGATGTTGAG
241   CAAGCTCTGC TGCCTGCTTA CCCGAAAGAG CTGAACTCTA CCCAGCCGTT CCAGCTGTTG
301   GAAACTTCAC CGCAGTTTAT TTATCAGGCA CAGAGCGGTC TGACCGGTCG TGATGGCCCG
361   GATAACCCGG CTAACGGCCC GCGTCCGATG TATAACGTTG AAAAAGACGC TTATGTGCTG
421   GCTGAAGGTC AAAACGAACT GCAGGTGCCG ATGACGTATA CCGACGCGGC AGGCAACACG
481   TTTACCAAAA CGTTTGTCCT GAAACGTGGT GATTACGCTG TCAACGTCAA CTACAACGTG
541   CAGAACGCTG GCGAGAAACC GCTGGAAATC TCCACCTTTG GTCAGTTGAA GCAATCCATC
601   ACTCTGCCAC CGCATCTCGA TACCGGAAGC AGCAACTTCG CACTGCACAC CTTCCGCGGC
661   GCGGCGTACT CCACGCCTGA CGAGAAGTAC GAGAAATACA AGTTCGATAC CATTGCCGAT
721   AACGAAAACC TGAACATCTC TTCGAAAGGT GGTTGGGTGG CAATGCTGCA ACAGTATTTC
781   GCGACGGCGT GGATCCCGCA TAACGACGGT ACCAACAACT TCTATACCGC TAATCTGGGT
841   AACGGCATCG CCGCTATCGG CTATAAATCT CAGCCGGTAC TGGTTCAGCC TGGTCAGACT
901   GGCGCGATGA ACAGCACCCT GTGGGTTGGC CCGGAAATCC AGGACAAAAT GGCAGCTGTT
961   GCTCCGCACC TGGATCTGAC CGTTGATTAC GGTTGGTTGT GGTTCATCTC TCAGCCGCTG
1021  TTCAAACTGC TGAAATGGAT CCATAGCTTT GTGGGTAACT GGGCTTCTC CATTATCATC
1081  ATCACCTTTA TCGTTCGTGG CATCATGTAC CCGCTGACCA AAGCGCAGTA CACCTCCATG
1141  GCGAAGATGC GTATGCTGCA GCCGAAGATT CAGGCAATGC GTGAGCGTCT GGGTGATGAC
1201  AAACAGCGTA TCAGCCAGGA AATGATGGCG CTGTACAAAG CTGAGAAGGT TAACCCGCTG
1261  GGCGGCTGCT TCCCGCTGCT GATCCAGATG CCAATCTTCC TGGCGTTGTA CTACATGCTG
1321  ATGGGTTCCG TTGAACTGCG TCAGGCACCG TTTGCACTGT GGATCCACGA CCTGTCGGCA
1381  CAGGACCCGT ACTACATCCT GCCGATCCTG ATGGGCGTAA CGATGTTCTT CATTCAGAAG
1441  ATGTCGCCGA CCACAGTGAC CGACCCGATG CAGCAGAAGA TCATGACCTT TATGCCGGTC
1501  ATCTTCACCG TGTTCTTCCT GTGGTTCCCG TCAGGTCTGG TGCTGTACTA TATCGTCAGC
1561  AACCTGGTAA CCATTATTCA GCAGCAGCTG ATTTACCGTG GTCTGGAAAA ACGTGGCCTG
1621  CATAGCCGCG AGAAGAAAAA ATCCAAGCTT GCGGCCGCAC TCGAGCACCA CCACCACCAC
1681  CACTGA
```

Protein sequence

```
  1    MDSQRNLLVI ALLFVSFMIW QAWEQDKNPQ PQAQQTTQTT TTAAGSAADQ GVPASGQGKL
 61    ISVKTDVLDL TINTRGGDVE QALLPAYPKE LNSTQPFQLL ETSPQFIYQA QSGLTGRDGP
121    DNPANGPRPM YNVEKDAYVL AEGQNELQVP MTYTDAAGNT FTKTFVLKRG DYAVNVNYNV
181    QNAGEKPLEI STFGQLKQSI TLPPHLDTGS SNFALHTFRG AAYSTPDEKY EKYKFDTIAD
241    NENLNISSKG GWVAMLQQYF ATAWIPHNDG TNNFYTANLG NGIAAIGYKS QPVLVQPGQT
301    GAMNSTLWVG PEIQDKMAAV APHLDLTVDY GWLWFISQPL FKLLKWIHSF VGNWGFSIII
361    ITFIVRGIMY PLTKAQYTSM AKMRMLQPKI QAMRERLGDD KQRISQEMMA LYKAEKVNPL
421    GGCFPLLIQM PIFLALYYML MGSVELRQAP FALWIHDLSA QDPYYILPIL MGVTMFFIQK
481    MSPTTVTDPM QQKIMTFMPV IFTVFFLWFP SGLVLYYIVS NLVTIIQQQL IYRGLEKRGL
541    HSREKKKSKL AAALEHHHHH H*
```

Number of amino acids: 561
Molecular weight: 63077.6
Theoretical pI: 7.77

His6 YidC

DNA sequence

```
   1   ATGTCGTACT ACCATCACCA TCACCATCAC GATTACGATA TCCCAACGAC CGAAAACCTG
  61   TATTTTCAGG GCGCCATGGG ATCCGGAATT CATATGGATT CGCAACGCAA TCTTTTAGTC
 121   ATCGCTTTGC TGTTCGTGTC TTTCATGATC TGGCAAGCCT GGGAGCAGGA TAAAAACCCG
 181   CAACCTCAGG CCCAACAGAC CACGCAGACA ACGACCACCG CAGCGGGTAG CGCCGCCGAC
 241   CAGGGCGTAC CGGCCAGTGG CCAGGGGAAA CTGATCTCGG TTAAGACCGA CGTGCTTGAT
 301   CTGACCATCA ACACCCGTGG TGGTGATGTT GAGCAAGCTC TGCTGCCTGC TTACCCGAAA
 361   GAGCTGAACT CTACCCAGCC GTTCCAGCTG TTGGAAACTT CACCGCAGTT TATTTATCAG
 421   GCACAGAGCG GTCTGACCGG TCGTGATGGC CCGGATAACC CGGCTAACGG CCCGCGTCCG
 481   ATGTATAACG TTGAAAAAGA CGCTTATGTG CTGGCTGAAG GTCAAAACGA ACTGCAGGTG
 541   CCGATGACGT ATACCGACGC GGCAGGCAAC ACGTTTACCA AAACGTTTGT CCTGAAACGT
 601   GGTGATTACG CTGTCAACGT CAACTACAAC GTGCAGAACG CTGGCGAGAA ACCGCTGGAA
 661   ATCTCCACCT TTGGTCAGTT GAAGCAATCC ATCACTCTGC CACCGCATCT CGATACCGGA
 721   AGCAGCAACT TCGCACTGCA CACCTTCCGC GGCGCGGCGT ACTCCACGCC TGACGAGAAG
 781   TACGAGAAAT ACAAGTTCGA TACCATTGCC GATAACGAAA ACCTGAACAT CTCTTCGAAA
 841   GGTGGTTGGG TGGCAATGCT GCAACAGTAT TTCGCGACGG CGTGGATCCC GCATAACGAC
 901   GGTACCAACA ACTTCTATAC CGCTAATCTG GGTAACGGCA TCGCCGCTAT CGGCTATAAA
 961   TCTCAGCCGG TACTGGTTCA GCCTGGTCAG ACTGGCGCGA TGAACAGCAC CCTGTGGGTT
1021   GGCCCGGAAA TCCAGGACAA AATGGCAGCT GTTGCTCCGC ACCTGGATCT GACCGTTGAT
1081   TACGGTTGGT TGTGGTTCAT CTCTCAGCCG CTGTTCAAAC TGCTGAAATG GATCCATAGC
1141   TTTGTGGGTA ACTGGGGCTT CTCCATTATC ATCATCACCT TTATCGTTCG TGGCATCATG
1201   TACCCGCTGA CCAAAGCGCA GTACACCTCC ATGGCGAAGA TGCGTATGCT GCAGCCGAAG
1261   ATTCAGGCAA TGCGTGAGCG TCTGGGTGAT GACAAACAGC GTATCAGCCA GGAAATGATG
1321   GCGCTGTACA AAGCTGAGAA GGTTAACCCG CTGGGCGGCT GCTTCCCGCT GCTGATCCAG
1381   ATGCCAATCT TCCTGGCGTT GTACTACATG CTGATGGGTT CCGTTGAACT GCGTCAGGCA
```

```
1441    CCGTTTGCAC TGTGGATCCA CGACCTGTCG GCACAGGACC CGTACTACAT CCTGCCGATC
1501    CTGATGGGCG TAACGATGTT CTTCATTCAG AAGATGTCGC CGACCACAGT GACCGACCCG
1561    ATGCAGCAGA AGATCATGAC CTTTATGCCG GTCATCTTCA CCGTGTTCTT CCTGTGGTTC
1621    CCGTCAGGTC TGGTGCTGTA CTATATCGTC AGCAACCTGG TAACCATTAT TCAGCAGCAG
1681    CTGATTTACC GTGGTCTGGA AAAACGTGGC CTGCATAGCC GCGAGAAGAA AAAATCCTGA
```

Protein sequence

```
1       MSYYHHHHHH DYDIPTTENL YFQGAMGSGI HMDSQRNLLV IALLFVSFMI WQAWEQDKNP
61      QPQAQQTTQT TTTAAGSAAD QGVPASGQGK LISVKTDVLD LTINTRGGDV EQALLPAYPK
121     ELNSTQPFQL LETSPQFIYQ AQSGLTGRDG PDNPANGPRP MYNVEKDAYV LAEGQNELQV
181     PMTYTDAAGN TFTKTFVLKR GDYAVNVNYN VQNAGEKPLE ISTFGQLKQS ITLPPHLDTG
241     SSNFALHTFR GAAYSTPDEK YEKYKFDTIA DNENLNISSK GGWVAMLQQY FATAWIPHND
301     GTNNFYTANL GNGIAAIGYK SQPVLVQPGQ TGAMNSTLWV GPEIQDKMAA VAPHLDLTVD
361     YGWLWFISQP LFKLLKWIHS FVGNWGFSII IITFIVRGIM YPLTKAQYTS MAKMRMLQPK
421     IQAMRERLGD DKQRISQEMM ALYKAEKVNP LGGCFPLLIQ MPIFLALYYM LMGSVELRQA
481     PFALWIHDLS AQDPYYILPI LMGVTMFFIQ KMSPTTVTDP MQQKIMTFMP VIFTVFFLWF
541     PSGLVLYYIV SNLVTIIQQQ LIYRGLEKRG LHSREKKKS*
```

Number of amino acids: 579
Molecular weight: 65236.9
Theoretical pI: 6.71

His6YidCΔc

DNA sequence

```
1       CAGACCATGT CGTACTACCA TCACCATCAC CATCACGATT ACGATATCCC AACGACCGAA
61      AACCTGTATT TCAGGGCGC CATGGGATCC GGAATTCATA TGGATTCGCA ACGCAATCTT
121     TTAGTCATCG CTTTGCTGTT CGTGTCTTTC ATGATCTGGC AAGCCTGGGA GCAGGATAAA
181     AACCCGCAAC CTCAGGCCCA ACAGACCACG CAGACAACGA CCACCGCAGC GGGTAGCGCC
241     GCCGACCAGG GCGTACCGGC CAGTGGCCAG GGGAAACTGA TCTCGGTTAA GACCGACGTG
301     CTTGATCTGA CCATCAACAC CCGTGGTGGT GATGTTGAGC AAGCTCTGCT GCCTGCTTAC
361     CCGAAAGAGC TGAACTCTAC CCAGCCGTTC CAGCTGTTGG AAACTTCACC GCAGTTTATT
421     TATCAGGCAC AGAGCGGTCT GACCGGTCGT GATGGCCCGG ATAACCCGGC TAACGGCCCG
481     CGTCCGATGT ATAACGTTGA AAAAGACGCT TATGTGCTGG CTGAAGGTCA AAACGAACTG
541     CAGGTGCCGA TGACGTATAC CGACGCGGCA GGCAACACGT TTACCAAAAC GTTTGTCCTG
601     AAACGTGGTG ATTACGCTGT CAACGTCAAC TACAACGTGC AGAACGCTGG CGAGAAACCG
661     CTGGAAATCT CCACCTTTGG TCAGTTGAAG CAATCCATCA CTCTGCCACC GCATCTCGAT
721     ACCGGAAGCA GCAACTTCGC ACTGCACACC TTCCGCGGCG CGGCGTACTC CACGCCTGAC
781     GAGAAGTACG AGAAATACAA GTTCGATACC ATTGCCGATA ACGAAAACCT GAACATCTCT
841     TCGAAAGGTG GTTGGGTGGC AATGCTGCAA CAGTATTTCG CGACGGCGTG GATCCCGCAT
901     AACGACGGTA CCAACAACTT CTATACCGCT AATCTGGGTA ACGGCATCGC CGCTATCGGC
961     TATAAATCTC AGCCGGTACT GGTTCAGCCT GGTCAGACTG GCGCGATGAA CAGCACCCTG
```

```
1021    TGGGTTGGCC CGGAAATCCA GGACAAAATG GCAGCTGTTG CTCCGCACCT GGATCTGACC
1081    GTTGATTACG GTTGGTTGTG GTTCATCTCT CAGCCGCTGT TCAAACTGCT GAAATGGATC
1141    CATAGCTTTG TGGGTAACTG GGGCTTCTCC ATTATCATCA TCACCTTTAT CGTTCGTGGC
1201    ATCATGTACC CGCTGACCAA AGCGCAGTAC ACCTCCATGG CGAAGATGCG TATGCTGCAG
1261    CCGAAGATTC AGGCAATGCG TGAGCGTCTG GGTGATGACA AACAGCGTAT CAGCCAGGAA
1321    ATGATGGCGC TGTACAAAGC TGAGAAGGTT AACCCGCTGG GCGGCTGCTT CCCGCTGCTG
1381    ATCCAGATGC CAATCTTCCT GGCGTTGTAC TACATGCTGA TGGGTTCCGT TGAACTGCGT
1441    CAGGCACCGT TTGCACTGTG GATCCACGAC CTGTCGGCAC AGGACCCGTA CTACATCCTG
1501    CCGATCCTGA TGGGCGTAAC GATGTTCTTC ATTCAGAAGA TGTCGCCGAC CACAGTGACC
1561    GACCCGATGC AGCAGAAGAT CATGACCTTT ATGCCGGTCA TCTTCACCGT GTTCTTCCTG
1621    TGGTTCCCGT CAGGTCTGGT GCTGTACTAT ATCGTCAGCA ACCTGGTAAC CATTATTCAG
1681    CAGCAGCTGA TTTACCGTGG TCTGTGA
```

Protein sequence

```
1      QTMSYYHHHH HHDYDIPTTE NLYFQGAMGS GIHMDSQRNL LVIALLFVSF MIWQAWEQDK
61     NPQPQAQQTT QTTTTAAGSA ADQGVPASGQ GKLISVKTDV LDLTINTRGG DVEQALLPAY
121    PKELNSTQPF QLLETSPQFI YQAQSGLTGR DGPDNPANGP RPMYNVEKDA YVLAEGQNEL
181    QVPMTYTDAA GNTFTKTFVL KRGDYAVNVN YNVQNAGEKP LEISTFGQLK QSITLPPHLD
241    TGSSNFALHT FRGAAYSTPD EKYEKYKFDT IADNENLNIS SKGGWVAMLQ QYFATAWIPH
301    NDGTNNFYTA NLGNGIAAIG YKSQPVLVQP GQTGAMNSTL WVGPEIQDKM AAVAPHLDLT
361    VDYGWLWFIS QPLFKLLKWI HSFVGNWGFS IIIITFIVRG IMYPLTKAQY TSMAKMRMLQ
421    PKIQAMRERL GDDKQRISQE MMALYKAEKV NPLGGCFPLL IQMPIFLALY YMLMGSVELR
481    QAPFALWIHD LSAQDPYYIL PILMGVTMFF IQKMSPTTVT DPMQQKIMTF MPVIFTVFFL
541    WFPSGLVLYY IVSNLVTIIQ QQLIYRGL*
```

Number of amino acids: 568
Molecular weight: 63901.3
Theoretical pI: 6.12

Oxa1$_{His6}$

DNA sequence

```
1      ATGAATTCGA CGGGCCCAAA TGCCAACGAT GTCTCGGAAA TCCAAACCCA GTTGCCTTCC
61     ATCGATGAAT TAACCTCTTC AGCTCCTTCT CTTTCCGCTT CTACTTCGGA CCTTATCGCT
121    AACACGACCC AAACAGTGGG CGAGTTGTCC TCCCATATAG GGTACTTAAA TAGCATTGGC
181    CTGGCCCAAA CCTGGTACTG GCCCTCGGAC ATTATCCAAC ACGTCTTGGA GGCCGTTCAT
241    GTTTACTCTG GGTTGCCTTG GTGGGGAACT ATCGCGGCCA CCACCATCCT CATTCGATGC
301    CTGATGTTTC CCCTCTATGT CAAGTCCTCT GATACTGTTG CTAGAAATTC CCATATCAAG
361    CCCGAGCTGG ACGCCTTGAA TAATAAGCTA ATGTCCACTA CAGATTTGCA ACAAGGTCAG
421    CTAGTCGCCA TGCAAAGGAA AAAACTGCTC TCCTCGCACG GCATTAAGAA CAGATGGCTG
481    GCCGCACCCA TGCTACAAAT TCCAATCGCC CTTGGGTTTT TCAACGCATT GAGACACATG
541    GCTAACTACC CAGTAGATGG GTTCGCTAAT CAAGGTGTCG CTTGGTTTAC AGACTTGACT
```

```
601   CAAGCAGACC CTTACTTAGG TTTGCAAGTA ATCACTGCCG CTGTGTTCAT CTCATTTACA
661   AGGCTGGGGG GTGAGACTGG TGCTCAACAA TTCAGTTCTC CCATGAAGCG TCTTTTCACT
721   ATTCTACCGA TCATTTCTAT ACCGGCCACA ATGAACTTAT CGTCCGCTGT GGTCCTCTAC
781   TTTGCCTTTA ATGGTGCCTT CTCCGTCCTA CAGACAATGA TTTTGAGAAA CAAATGGGTT
841   CGTTCGAAAC TGAAGATAAC AGAAGTAGCT AAACCAAGGA CTCCTATCGC TGGCGCTTCC
901   CCCACAGAGA ACATGGGCAT CTTCCAATCA TTAAAACATA ACATTCAAAA GGCAAGAGAT
961   CAGGCGGAAA GAAGGCAATT GATGCAAGAT AATGAGAAGA AGTTACAAGA AAGCTTCAAG
1021  GAGAAGAGGC AGAATTCCAA AATCAAAATT GTTCACAAAT CAAACTTCAT TAATAACAAA
1081  AAAGGATCAC TCGAGCACCA CCACCACCAC CACTGA
```

Protein sequence

```
1     MNSTGPNAND VSEIQTQLPS IDELTSSAPS LSASTSDLIA NTTQTVGELS SHIGYLNSIG
61    LAQTWYWPSD IIQHVLEAVH VYSGLPWWGT IAATTILIRC LMFPLYVKSS DTVARNSHIK
121   PELDALNNKL MSTTDLQQGQ LVAMQRKKLL SSHGIKNRWL AAPMLQIPIA LGFFNALRHM
181   ANYPVDGFAN QGVAWFTDLT QADPYLGLQV ITAAVFISFT RLGGETGAQQ FSSPMKRLFT
241   ILPIISIPAT MNLSSAVVLY FAFNGAFSVL QTMILRNKWV RSKLKITEVA KPRTPIAGAS
301   PTENMGIFQS LKHNIQKARD QAERRQLMQD NEKKLQESFK EKRQNSKIKI VHKSNFINNK
361   KGSLEHHHHH H*
```

Number of amino acids: 371
Molecular weight: 41340.5
Theoretical pI: 9.66

F_0c first TMH nascent chain construct (if not specified, RNCs display this chain)

DNA sequence

```
1     ATGGCTAGCT GGAGCCACCC GCAGTTCGAA AAAGGCGCCA TGACCGGTTG GAGCCACCCG
61    CAGTTCGAAA AACGGTCCGC CGGGTCCTGG AGCCACCCGC AGTTCGAAAA ACTGCAGATG
121   GAAAACCTGA ATATGGATCT GCTGTACATG GCTGCCGCTG TGATGATGGG TCTGGCGGCA
181   ATCGGTGCTG CGATCGGTAT CGGCATCCTC GGGGGTAAAT TCCTGGAAGG CGCAGCGCGT
241   CAACCTGATC TGATTGATAT CTCTGAAAAG GGTTATCGCA TTGATTATGC GCATTTTACC
301   CCACAAGCAA AATTCAGCAC GCCCGTCTGG ATAAGCCAGG CGCAAGGCAT CCGTGCTGGC
361   CCTCAACGCC TCAGCTTCAT GATGATGATG ATGATGTGA
```

Protein sequence

```
1     MASWSHPQFE KGAMTGWSHP QFEKRSAGSW SHPQFEKLQM ENLNMDLLYM AAAVMMGLAA
61    IGAAIGIGIL GGKFLEGAAR QPDLIDISEK GYRIDYAHFT PQAKFSTPVW ISQAQGIRAG
121   PQRLSFMMMM MM*
```

Number of amino acids: 132
Molecular weight: 14671.1
Theoretical pI: 8.09

F_0c full length 2TMH nascent chain construct

DNA sequence

```
1     ATGGCTAGCT GGAGCCACCC GCAGTTCGAA AAAGGCGCCA TGACCGGTTG GAGCCACCCG
61    CAGTTCGAAA AACGGTCCGC CGGGTCCTGG AGCCACCCGC AGTTCGAAAA ACTGCAGATG
121   GAAAACCTGA ATATGGATCT GCTGTACATG GCTGCCGCTG TGATGATGGG TCTGGCGGCA
181   ATCGGTGCTG CGATCGGTAT CGGCATCCTC GGGGGTAAAT TCCTGGAAGG CGCAGCGCGT
241   CAACCTGATC TGATTCCTCT GCTGCGTACT CAGTTCTTTA TCGTTATGGG TCTGGTGGAT
301   GCTATCCCGA TGATCGCTGT AGGTCTGGGT CTGTACGTGA TGTTCGCTGT CGCGGATATC
361   TCTGAAAAGG GTTATCGCAT TGATTATGCG CATTTTACCC CACAAGCAAA ATTCAGCACG
421   CCCGTCTGGA TAAGCCAGGC GCAAGGCATC CGTGCTGGCC CTCAACGCCT CAGCTTCATG
481   ATGATGATGA TGATGTGA
```

Protein sequence

```
1     MASWSHPQFE KGAMTGWSHP QFEKRSAGSW SHPQFEKLQM ENLNMDLLYM AAAVMMGLAA
61    IGAAIGIGIL GGKFLEGAAR QPDLIPLLRT QFFIVMGLVD AIPMIAVGLG LYVMFAVADI
121   SEKGYRIDYA HFTPQAKFST PVWISQAQGI RAGPQRLSFM MMMM*
```

Number of amino acids: 165
Molecular weight: 18220.5
Theoretical pI: 8.06

7 REFERENCES

Abramoff, M. D., P. J. Magelhaes, et al. (2004). "Image Processing with ImageJ." Biophotonics International **11**(7): 36-42.

Antoun, A., M. Y. Pavlov, et al. (2006). "How initiation factors maximize the accuracy of tRNA selection in initiation of bacterial protein synthesis." Mol Cell **23**(2): 183-93.

Ban, N., P. Nissen, et al. (2000). "The complete atomic structure of the large ribosomal subunit at 2.4 A resolution." Science **289**(5481): 905-20.

Beck, K., G. Eisner, et al. (2001). "YidC, an assembly site for polytopic Escherichia coli membrane proteins located in immediate proximity to the SecYE translocon and lipids." EMBO Rep **2**(8): 709-14.

Becker, T., S. Bhushan, et al. (2009). "Structure of monomeric yeast and mammalian Sec61 complexes interacting with the translating ribosome." Science **326**(5958): 1369-73.

Bessonneau, P., V. Besson, et al. (2002). "The SecYEG preprotein translocation channel is a conformationally dynamic and dimeric structure." Embo J **21**(5): 995-1003.

Bonnefoy, N., F. Chalvet, et al. (1994). "OXA1, a Saccharomyces cerevisiae nuclear gene whose sequence is conserved from prokaryotes to eukaryotes controls cytochrome oxidase biogenesis." J Mol Biol **239**(2): 201-12.

Bostina, M., B. Mohsin, et al. (2005). "Atomic model of the E. coli membrane-bound protein translocation complex SecYEG." J Mol Biol **352**(5): 1035-43.

Breyton, C., W. Haase, et al. (2002). "Three-dimensional structure of the bacterial protein-translocation complex SecYEG." Nature **418**(6898): 662-5.

Brundage, L., J. P. Hendrick, et al. (1990). "The purified E. coli integral membrane protein SecY/E is sufficient for reconstitution of SecA-dependent precursor protein translocation." Cell **62**(4): 649-57.

Cannon, K. S., E. Or, et al. (2005). "Disulfide bridge formation between SecY and a translocating polypeptide localizes the translocation pore to the center of SecY." J Cell Biol **169**(2): 219-25.

Chen, M., J. C. Samuelson, et al. (2002). "Direct interaction of YidC with the Sec-independent Pf3 coat protein during its membrane protein insertion." J Biol Chem **277**(10): 7670-5.

Clemons, W. M., Jr., J. F. Menetret, et al. (2004). "Structural insight into the protein translocation channel." Curr Opin Struct Biol **14**(4): 390-6.

DeLano WL (2002). The PyMOL Molecular Graphics System. Palo Alto, CA, USA, DeLano Scientific.

Dong, Y., S. R. Palmer, et al. (2008). "Functional overlap but lack of complete cross-complementation of Streptococcus mutans and Escherichia coli YidC orthologs." J Bacteriol **190**(7): 2458-69.

Driessen, A. J. (1994). "How proteins cross the bacterial cytoplasmic membrane." J Membr Biol **142**(2): 145-59.

Dubochet, J., M. Adrian, et al. (1988). "Cryo-electron microscopy of vitrified specimens." Q Rev Biophys **21**(2): 129-228.

Facey, S. J., S. A. Neugebauer, et al. (2007). "The mechanosensitive channel protein MscL is targeted by the SRP to the novel YidC membrane insertion pathway of Escherichia coli." J Mol Biol **365**(4): 995-1004.

Falk, S., S. Ravaud, et al. "The C terminus of the Alb3 membrane insertase recruits cpSRP43 to the thylakoid membrane." J Biol Chem **285**(8): 5954-62.

Fancy, D. A. and T. Kodadek (1999). "Chemistry for the analysis of protein-protein interactions: rapid and efficient cross-linking triggered by long wavelength light." Proc Natl Acad Sci U S A **96**(11): 6020-4.

Frank, J., M. Radermacher, et al. (1996). "SPIDER and WEB: processing and visualization of images in 3D electron microscopy and related fields." J Struct Biol **116**(1): 190-9.

Froderberg, L., E. N. Houben, et al. (2004). "Targeting and translocation of two lipoproteins in Escherichia coli via the SRP/Sec/YidC pathway." J Biol Chem **279**(30): 31026-32.

Glick, B. S. and G. Von Heijne (1996). "Saccharomyces cerevisiae mitochondria lack a bacterial-type sec machinery." Protein Sci **5**(12): 2651-2.

Hardesty, B. and G. Kramer (2001). "Folding of a nascent peptide on the ribosome." Prog Nucleic Acid Res Mol Biol **66**: 41-66.

Hirashima, A. and A. Kaji (1973). "Role of elongation factor G and a protein factor on the release of ribosomes from messenger ribonucleic acid." J Biol Chem **248**(21): 7580-7.

Jia, L., M. Dienhart, et al. (2003). "Yeast Oxa1 interacts with mitochondrial ribosomes: the importance of the C-terminal region of Oxa1." EMBO J **22**(24): 6438-47.

Jia, L., M. K. Dienhart, et al. (2007). "Oxa1 directly interacts with Atp9 and mediates its assembly into the mitochondrial F1Fo-ATP synthase complex." Mol Biol Cell **18**(5): 1897-908.

Jiang, F., M. Chen, et al. (2003). "Defining the regions of Escherichia coli YidC that contribute to activity." J Biol Chem **278**(49): 48965-72.

Jiang, F., L. Yi, et al. (2002). "Chloroplast YidC homolog Albino3 can functionally complement the bacterial YidC depletion strain and promote membrane insertion of both bacterial and chloroplast thylakoid proteins." J Biol Chem **277**(22): 19281-8.

Keenan, R. J., D. M. Freymann, et al. (2001). "The signal recognition particle." Annu Rev Biochem **70**: 755-75.

Kiefer, D. and A. Kuhn (2007). "YidC as an essential and multifunctional component in membrane protein assembly." Int Rev Cytol **259**: 113-38.

Klenner, C., J. Yuan, et al. (2008). "The Pf3 coat protein contacts TM1 and TM3 of YidC during membrane biogenesis." FEBS Lett **582**(29): 3967-72.

Kohler, R., D. Boehringer, et al. (2009). "YidC and Oxa1 form dimeric insertion pores on the translating ribosome." Mol Cell **34**(3): 344-53.

LeBarron, J., R. A. Grassucci, et al. (2008). "Exploration of parameters in cryo-EM leading to an improved density map of the E. coli ribosome." J Struct Biol **164**(1): 24-32.

Lotz, M., W. Haase, et al. (2008). "Projection structure of yidC: a conserved mediator of membrane protein assembly." J Mol Biol **375**(4): 901-7.

Ludtke, S. J., P. R. Baldwin, et al. (1999). "EMAN: semiautomated software for high-resolution single-particle reconstructions." J Struct Biol **128**(1): 82-97.

Luirink, J., T. Samuelsson, et al. (2001). "YidC/Oxa1p/Alb3: evolutionarily conserved mediators of membrane protein assembly." FEBS Lett **501**(1): 1-5.

Luirink, J., G. von Heijne, et al. (2005). "Biogenesis of inner membrane proteins in Escherichia coli." Annu Rev Microbiol **59**: 329-55.

Mathieu, L., M. Bourens, et al. (2010). "A mutational analysis reveals new functional interactions between domains of the Oxa1 protein in Saccharomyces cerevisiae." Mol Microbiol **75**(2): 474-88.

Menetret, J. F., R. S. Hegde, et al. (2008). "Single copies of Sec61 and TRAP associate with a nontranslating mammalian ribosome." Structure **16**(7): 1126-37.

Menetret, J. F., J. Schaletzky, et al. (2007). "Ribosome binding of a single copy of the SecY complex: implications for protein translocation." Mol Cell **28**(6): 1083-92.

Merz, F., D. Boehringer, et al. (2008). "Molecular mechanism and structure of Trigger Factor bound to the translating ribosome." EMBO J **27**(11): 1622-32.

Mitra, K., C. Schaffitzel, et al. (2005). "Structure of the E. coli protein-conducting channel bound to a translating ribosome." Nature **438**(7066): 318-24.

Moazed, D. and H. F. Noller (1989). "Intermediate states in the movement of transfer RNA in the ribosome." Nature **342**(6246): 142-8.

Nagamori, S., I. N. Smirnova, et al. (2004). "Role of YidC in folding of polytopic membrane proteins." J Cell Biol **165**(1): 53-62.

Nissen, P., J. Hansen, et al. (2000). "The structural basis of ribosome activity in peptide bond synthesis." Science **289**(5481): 920-30.

Nouwen, N. and A. J. Driessen (2002). "SecDFyajC forms a heterotetrameric complex with YidC." Mol Microbiol **44**(5): 1397-405.

Oliver, D. C. and M. Paetzel (2008). "Crystal structure of the major periplasmic domain of the bacterial membrane protein assembly facilitator YidC." J Biol Chem **283**(8): 5208-16.

Osborne, A. R. and T. A. Rapoport (2007). "Protein translocation is mediated by oligomers of the SecY complex with one SecY copy forming the channel." Cell **129**(1): 97-110.

Pop, O. I., Z. Soprova, et al. (2009). "YidC is required for the assembly of the MscL homopentameric pore." Febs J **276**(17): 4891-9.

Rapoport, T. A. (2007). "Protein translocation across the eukaryotic endoplasmic reticulum and bacterial plasma membranes." Nature **450**(7170): 663-9.

Rapoport, T. A. (2008). "Protein transport across the endoplasmic reticulum membrane." FEBS J **275**(18): 4471-8.

Ravaud, S., G. Stjepanovic, et al. (2008). "The crystal structure of the periplasmic domain of the Escherichia coli membrane protein insertase YidC contains a substrate binding cleft." J Biol Chem **283**(14): 9350-8.

Robson, A., B. Carr, et al. (2009). "Synthetic peptides identify a second periplasmic site for the plug of the SecYEG protein translocation complex." FEBS Lett **583**(1): 207-12.

Samuelson, J. C., M. Chen, et al. (2000). "YidC mediates membrane protein insertion in bacteria." Nature **406**(6796): 637-41.

Sander, B., M. M. Golas, et al. (2003). "Automatic CTF correction for single particles based upon multivariate statistical analysis of individual power spectra." J Struct Biol **142**(3): 392-401.

Schaffitzel, C. and N. Ban (2007). "Generation of ribosome nascent chain complexes for structural and functional studies." J Struct Biol **158**(3): 463-71.

Schaffitzel, C., M. Oswald, et al. (2006). "Structure of the E. coli signal recognition particle bound to a translating ribosome." Nature **444**(7118): 503-6.

Schuwirth, B. S., M. A. Borovinskaya, et al. (2005). "Structures of the bacterial ribosome at 3.5 A resolution." Science **310**(5749): 827-34.

Scotti, P. A., M. L. Urbanus, et al. (2000). "YidC, the Escherichia coli homologue of mitochondrial Oxa1p, is a component of the Sec translocase." EMBO J **19**(4): 542-9.

Selmer, M., C. M. Dunham, et al. (2006). "Structure of the 70S ribosome complexed with mRNA and tRNA." Science **313**(5795): 1935-42.

Serek, J., G. Bauer-Manz, et al. (2004). "Escherichia coli YidC is a membrane insertase for Sec-independent proteins." EMBO J **23**(2): 294-301.

Shine, J. and L. Dalgarno (1974). "The 3'-terminal sequence of Escherichia coli 16S ribosomal RNA: complementarity to nonsense triplets and ribosome binding sites." Proc Natl Acad Sci U S A **71**(4): 1342-6.

Szyrach, G., M. Ott, et al. (2003). "Ribosome binding to the Oxa1 complex facilitates co-translational protein insertion in mitochondria." EMBO J **22**(24): 6448-57.

Urbanus, M. L., L. Froderberg, et al. (2002). "Targeting, insertion, and localization of Escherichia coli YidC." J Biol Chem **277**(15): 12718-23.

Urbanus, M. L., P. A. Scotti, et al. (2001). "Sec-dependent membrane protein insertion: sequential interaction of nascent FtsQ with SecY and YidC." EMBO Rep **2**(6): 524-9.

Valle, M., J. Sengupta, et al. (2002). "Cryo-EM reveals an active role for aminoacyl-tRNA in the accommodation process." EMBO J **21**(13): 3557-67.

Valle, M., A. Zavialov, et al. (2003). "Incorporation of aminoacyl-tRNA into the ribosome as seen by cryo-electron microscopy." Nat Struct Biol **10**(11): 899-906.

van Bloois, E., H. L. Dekker, et al. (2008). "Detection of cross-links between FtsH, YidC, HflK/C suggests a linked role for these proteins in quality control upon insertion of bacterial inner membrane proteins." FEBS Lett **582**(10): 1419-24.

van Bloois, E., G. Koningstein, et al. (2007). "Saccharomyces cerevisiae Cox18 complements the essential Sec-independent function of Escherichia coli YidC." FEBS J **274**(21): 5704-13.

van Bloois, E., S. Nagamori, et al. (2005). "The Sec-independent function of Escherichia coli YidC is evolutionary-conserved and essential." J Biol Chem **280**(13): 12996-3003.

Van den Berg, B., W. M. Clemons, Jr., et al. (2004). "X-ray structure of a protein-conducting channel." Nature **427**(6969): 36-44.

van der Laan, M., P. Bechtluft, et al. (2004). "F1F0 ATP synthase subunit c is a substrate of the novel YidC pathway for membrane protein biogenesis." J Cell Biol **165**(2): 213-22.

van der Laan, M., E. N. Houben, et al. (2001). "Reconstitution of Sec-dependent membrane protein insertion: nascent FtsQ interacts with YidC in a SecYEG-dependent manner." EMBO Rep **2**(6): 519-23.

van der Laan, M., M. L. Urbanus, et al. (2003). "A conserved function of YidC in the biogenesis of respiratory chain complexes." Proc Natl Acad Sci U S A **100**(10): 5801-6.

van Heel, M., G. Harauz, et al. (1996). "A new generation of the IMAGIC image processing system." J Struct Biol **116**(1): 17-24.

Vinayagam, A., G. Pugalenthi, et al. (2004). "DSDBASE: a consortium of native and modelled disulphide bonds in proteins." Nucleic Acids Res **32**(Database issue): D200-2.

Wimberly, B. T., D. E. Brodersen, et al. (2000). "Structure of the 30S ribosomal subunit." Nature **407**(6802): 327-39.

Winterfeld, S., N. Imhof, et al. (2009). "Substrate-induced conformational change of the Escherichia coli membrane insertase YidC." Biochemistry **48**(28): 6684-91.

Wittig, I., H. P. Braun, et al. (2006). "Blue native PAGE." Nat Protoc **1**(1): 418-28.

Youngman, E. M., M. E. McDonald, et al. (2008). "Peptide release on the ribosome: mechanism and implications for translational control." Annu Rev Microbiol **62**: 353-73.

Yu, Z., G. Koningstein, et al. (2008). "The Conserved Third Transmembrane Segment of YidC Contacts Nascent Escherichia coli Inner Membrane Proteins." J Biol Chem **283**(50): 34635-42.

Yuan, J., G. J. Phillips, et al. (2007). "Isolation of cold-sensitive yidC mutants provides insights into the substrate profile of the YidC insertase and the importance of transmembrane 3 in YidC function." J Bacteriol **189**(24): 8961-72.

Glossary

aa	Amino acid
A. thaliana	*Arabidopsis thaliana*
ADA	N-(Carbamoylmethyl)iminodiacetic acid
Amp	Ampicillin
APS	Ammonium persulfate
ATP	Adenosine 5'triphosphate
B. G.	Basil Greber
bp	Base pair(s)
BSA	Bovine serum albumine
C.S.	Christiane Schaffitzel
CTAB	Cetyltrimethylammonium bromide
CV	Column volume
D. B.	Daniel Boehringer
DDM	n-Dodecyl-β-maltopyranoside
DM	n-Decyl-β-maltopyranoside
DMSO	Dimethylsulfoxide
DNA	Deoxyribonucleic acid
DTT	1,4-Dithiothreitol
E. coli	*Escherichia coli*
EM	Electron microscopy
EtBr	Ethidium bromide
H. marismortui	*Haloarcula marismortui*
H6-tag	Hexahistidine-tag
I. C.	Ian Collinson
IgG	Immunoglobulin G
IPTG	Isopropyl-1-thio-β-D-galactopyranoside
Kan	Kanamycin sulfate
kDa	Kilodalton
LB	Luria-Bertani
MDa	Megadalton
min	Minutes
M. Peter	Prof. Matthias Peter, Institute of Biochemistry, ETHZ

MW	Molecular weight (g/mol)
N. D.	Nadine Dörwald
NiNTA	Nickel-nitrilotriacetic acid
o/n	Over night
ODx	Optical density measured at x nm
OAc	Acetate anion
PAGE	Polyacrylamide gel electrophoresis
PCR	Polymerase chain reaction
PDF	Peptide deformylase
PMSF	Phenylmethyl sulfonyl fluoride
RNA	Ribonucleic acid
rpm	Rounds per minute
RT	Room temperature
Ru(bpy)$_3$	Tris-bipyridylruthenium(II)
S. cerevisiae	*Saccharomyces cerevisiae*
sec	Seconds
SDS	Sodium dodecyl sulfate
SRP	Signal recognition particle
TB	Terrific broth
TEMED	N,N,N',N'-Tetramethylethylenediamine
TF	Trigger factor
U	Unit
UV	Ultra violet
v/v	Volume per volume
vs.	Versus
w/v	Weight per volume
wt	Wild type
X-gal	(5-bromo-4-chloro-3-indolyl-beta-D-galacto-pyranoside)

i want morebooks!

Buy your books fast and straightforward online - at one of world's fastest growing online book stores! Environmentally sound due to Print-on-Demand technologies.

Buy your books online at
www.get-morebooks.com

Kaufen Sie Ihre Bücher schnell und unkompliziert online – auf einer der am schnellsten wachsenden Buchhandelsplattformen weltweit! Dank Print-On-Demand umwelt- und ressourcenschonend produziert.

Bücher schneller online kaufen
www.morebooks.de

VDM Verlagsservicegesellschaft mbH
Heinrich-Böcking-Str. 6-8   Telefon: +49 681 3720 174   info@vdm-vsg.de
D - 66121 Saarbrücken   Telefax: +49 681 3720 1749   www.vdm-vsg.de

Printed by Books on Demand GmbH, Norderstedt / Germany